Health and Safety in Construction Design

HEALTH AND SAFETY IN CONSTRUCTION DESIGN

Brian Thorpe

GOWER

Published by
Gower Publishing Limited
Gower House
Croft Road
Aldershot
Hants GU11 3HR
England

Gower Publishing Company
Suite 420
101 Cherry Street
Burlington
VT 05401-4405
USA

Brian Thorpe has asserted his right under the Copyright, Designs and Patents Act, 1988 to be identified as the author of this work.

British Library Cataloguing in Publication Data
Thorpe, Brian
  Health and safety in construction design. – (the leading construction series)
  1. Engineering design – Safety regulations – Great Britain
  2. Construction industry – Great Britain – Safety measures
  3. Construction industry – Law and legislation – Great Britain – Health and
  hygiene
  I. Title
  624'.0289

  ISBN 0 566 08670 0

Library of Congress Cataloging-in-Publication Data
Thorpe, Brian
  Health and safety in construction design / Brian Thorpe.
    p. cm. -- (The leading construction series)
  ISBN 0-566-08670-0
  1. Building sites--Safety measures. 2. Construction Industry--Safety measures. I. Title.
  II. Series

  TH443.T48 2005
  690'.22--dc22

                                                                              2004059798

Typeset by IML Typographers, Birkenhead, Merseyside
Printed and bound in Great Britain by MPG Books Ltd, Bodmin, Cornwall

Contents

List of Figures	*vii*
Abbreviations	*ix*
Preface	*xi*
Introduction	*xv*

Chapter 1 – Regulatory Requirements   **1**

Background   2

Levels of importance of various health and safety-related publications   3

    The Health and Safety at Work etc. Act (HASAWA)   3

    Regulations   4

    Codes of practice   4

    Guidance notes   5

A closer look at a selection of the key regulations   5

    The Management of Health and Safety at Work Regulations 1999 (MHSWR)   5

    Control of Substances Hazardous to Health (COSHH)   8

    Manual Handling Operations Regulations 1992   10

    Construction (Health, Safety and Welfare) Regulations 1996 (CHSWR)   10

    Construction (Design and Management) Regulations 1994 (CDM)   15

    Summary   23

The role and responsibilities of the health and safety inspector   23

Chapter 2 – The Design Process   **25**

The multiple responsibilities of the designer   26

Perceived shortcomings in design in relation to health and safety in construction   26

The indicated solution: a health and safety culture   29

Defining policy   30

An attainment strategy   31

    Step 1: Implement a training programme   31

    Step 2: Establish a health and safety database   31

    Step 3: Plan and resource health and safety requirements into the design programmes   33

    Step 4: Integrate health and safety requirements into the management system   34

    Step 5: Communicate with others   35

An application methodology   36

Chapter 3 – Application in Design   **39**

The designer's responsibilities and limitations   40

    What is a designer?   40

    The designer's duties and limitations   41

Hazards and risks in relation to design   42

    What are hazards and risks?   42

    Common types of hazard and associated risks   42

Setting up a health and safety-related database   51

Establishing site-related information   52
Application regulations and standards   54
Establishing information on types of hazard and risks   54
Precautions   55
Additional guideline   55
Expanding database information using peer inputs   56
Extracting and developing information   56
Brainstorming   57
Using your database to best advantage   57
Performance inhibitors   58
The benefits of teamworking and cooperation   59
Planning, resourcing and giving transparency to health and safety requirements   59
Assessing risks   60
Some useful techniques for assessing risks   67
The numerical approach   67
Paired comparisons   68
An alternative approach   69
Reviewing health and safety performance   72
Getting the right performance framework   72
Measuring performance   73
Relevant publications   74
Regulations   74
Further reading on hazards   74

Chapter 4 – An Integrated Management System Approach   **77**
Incorporating health and safety requirements into the design office system   78
Managing health and safety on a project-by-project basis   80
Providing visibility of health and safety measure pre-considerations, planning,
    resourcing, implementation and review   85
Achieving performance feedback and continuing improvement   86
Audits   86
Management reviews   87
Key performance indicators   87

Chapter 5 – Conclusion   **89**

Index   *93*

List of Figures

2.1   *Simple cause and effect diagram, showing some (not all) typical root causes of substandard design output*   28

2.2   *Typical example of how a construction can cause post-construction problems for others*   32

2.3   *Further example of how a construction can cause post-construction problems for others*   33

3.1   *Safe and unsafe site entrance design*   50

3.2   *Eliminating the necessity for working at height*   51

3.3   *Reducing the risks to others in lifting heavy loads*   51

3.4   *Proposed layout for a site-related information checklist*   53

3.5   *Hypothetical detail quality plan, based on the sequence of health and safety-related activities listed in Chapter 2*   61

3.6   *Paired comparisons diagram*   68

3.7   *Risk assessment question and answer form*   70

3.8   *Specimen completed risk assessment question and answer form*   71

4.1   *Typical document-type inputs into a project quality plan*   82

4.2   *Typical project quality plan inclusions*   82

ACOP	Approved Code of Practice
CDM	Construction (Design & Management) Regulations
CHSWR	Construction (Health, Safety & Welfare) Regulations
CIC	Construction Industry Council
CIRIA	Construction Industry Research and Information Association
COSHH	Control of Substances Hazardous to Health
DOP	detail quality plan
H & S	Health & Safety
HASAWA	Health & Safety at Work Act
HSE	Health & Safety Executive
KPI	Key Performance Indicator
MHSWE	Management of Health & Safety at Work Regulations
PQP	project quality plan
QMS	Quality Management System
RIBA	Royal Institute of British Architects

By its very nature, the construction industry is prone to hazards and associated risks for those engaged in both the actual construction and post-construction (for example, maintenance and cleaning) activities. Despite there being a wide variety of published information relating to good health and safety practice, such as regulations, ACOPs, guidance bulletins and so on, fatality and injury levels still persist, with little signs of real sustained improvement. Why is this so and what else can be done to improve matters?

This book aims to come to grips with this situation. It takes a holistic view embracing all the main parties involved in the construction process, although the main emphasis is placed upon the pre-construction stages of design, planning and training, plus the need for improved inter-discipline communications and health and safety-related awareness.

Throughout the book there is an emphasis on people. It should never be forgotten that, although we may have numerous regulations and guidelines, these in themselves cannot make things happen. Only people can do that. Unless people are really aware of their duties and obligations, understand what is expected of them personally, have been trained appropriately and possess the individual commitment to doing the right things, matters are not likely to improve significantly.

In other words, what we need to aim for is a culture of both individual and collective responsibility towards health and safety. Somehow the perceptions of many need to be widened from seeing health and safety matters as something which can be left to others (such as the safety officer or principal contractor) to something which is important to everyone at every level and as much an integral part of design and construction thinking as function, constructability, maintenance and cost.

The failure to meet health and safety requirements needs to be seen in terms of the reputation of both the industry as a whole and the companies concerned. An organization associated with adverse accident publicity is likely to project a very poor image which competitors can, and almost certainly will, exploit, leading to a lowering of market confidence, possible removal from lists of 'preferred suppliers' and a consequential loss of business opportunities.

You should not lose sight of the fact that among the selection criteria used to establish lists of preferred suppliers for goods or services, will increasingly be an ability to meet health and safety expectations, and rightly so. The Construction (Design and Management) regulations (CDM) require no less. During 2002–2003 the HSE through its Construction Division has launched a number of initiatives relating to construction design. These have included such things as:

- the setting up of a pilot 'design self-help group'
- the holding of 'design awareness days'
- meetings with regional CDM duty holders support groups to discuss best practice in 'design risk assessments'
- the carrying out of a number of 'in depth designer audits.

Probably the most publicized findings reported, however, were those concerning some 123 major projects, when HSE inspectors held meetings with invited designers from the north of England and Scotland in March 2003. While the main emphasis of the discussions concerned falls from heights – the biggest single cause of fatalities and accidents – some worrying revelations emerged:

- About a third of the designers had little or no knowledge of their responsibilities under the CDM regulations, with a mere 8 per cent saying that they had received training in the regulations.
- Many of the written design risk assessments were seen to be of poor quality and added little, if anything, to the safety of the construction process.
- Designers were often abdicating their responsibilities for reducing risks in relation to working at height and leaving it to the principal contractor to develop solutions.
- Contractors were struggling to control risks, which could easily have been eliminated or considerably reduced by good design.

Clearly, much needs to be done to improve matters.

You cannot begin to cure a problem unless you know what the root of the problem really is. Being reactive and rectifying things after they have occurred is not really effective unless we learn from mistakes, identify their root causes, eliminate them and spread the awareness to others.

This, of course, is the purpose of the many excellent bulletins issued within the industry, but, unfortunately, the circulation of sound advisory information is in itself no guarantee that its message has been properly cascaded down to those affected, through proper management-led, in-house staff training, coupled to an assessment of post-training understanding.

However, in the spring of 2004, the HSE carried out a similar exercise in the same geographical area and involving some 122 organizations, to the one undertaken during 2003. This showed a measurable improvement compared to the previous findings.

- Of those organizations visited, 47 per cent were able to demonstrate good or adequate training in the CDM regulations (previously 8 per cent).

- Of those organizations visited, 60 per cent demonstrated adequate knowledge of CDM requirements (previously 33 per cent).

The latter improvement would in some degree, it is presumed, be a consequence of the former. If these improved findings are typical of what is happening across the wider picture, this is encouraging and hopefully a reflection upon the efforts of the HSE and others to address problems through initiatives such as those mentioned above.

There is, however, no room for complacency and, despite the latest findings, there is still much to be done.

Another tool for improving health and safety is the risk assessment, but this has also been brought into question, including the methodology used, the subjectivity of decisions reached, the depth of analysis and the way in which findings are shared. It is not just knowledge and training that will achieve the health and safety objective that we strive to attain, it is the commitment of individuals to applying that knowledge and training in their day-to-day design work that will make the difference and ultimately result in the only real measure of progress that matters – namely, a reduction in the number of deaths and injuries within the industry.

When dealing with an industry involving many disciplines, wide variations in people's knowledge and experience and often little or no real knowledge of the duties and obligations of those who interface with us, solutions to problems are not straightforward. Nevertheless, on the basis that any problem is an opportunity to overcome a difficulty, you should not lose hope. Things can be done and, hopefully, this book will provide some help.

This book has been written in response to a need identified by a number of professional bodies associated with the construction industry. Although it discusses the wider implications of health and safety in construction, it focuses primarily on the design function (which essentially determines everything subsequent) and its significance in relation to health and safety during the actual construction stage and the post-construction areas of safety in maintenance and cleaning. How successful or otherwise this attempt has been only you, the reader, can decide.

I wish you an interesting and helpful read.

Brian Thorpe
2005

This book consists of five chapters.

Chapter 1 looks at some of the statutory and regulatory requirements to which the industry is subject, including the Health and Safety at Work etc. Act 1974 and a number of key regulations.

These are explored, first, using a general perspective and then in a slightly more detailed manner, with the objective of equating their main implications for site contractors and others to a perceived need for designer awareness and possible action. This comparison provides useful guidance concerning what should be considered for inclusion in any health and safety-related design database. The legal hierarchy of the above documents (that is, Acts, Regulations, ACOPs and so on) are touched upon, as are the designer's responsibilities concerning compliance with them.

The chapter then continues with an in-depth look at the CDM regulations with particular emphasis on Regulation 13 and the role of the designer. It concludes with a short summary of the duties and authority of the health and safety inspectors.

Chapter 2 takes a broad-based look at the design process and why health and safety in construction and beyond needs to be an integral part of this process. The indications arising from the findings of various audits and surveys, including those carried out by the HSE, are considered and the importance of establishing an organization-wide health and safety culture is discussed.

We then consider how this may be brought about through a clear health and safety policy, authorized by senior management and supported by a strategy to ensure its implementation, embracing such things as:

- appropriate staff training
- establishing a health and safety database
- planning and resourcing health and safety measures into design programmes
- integrating health and safety requirements into your management system
- communicating with others (who, when, why, how)
- monitoring performance, feedback and continuing improvement.

All the above are dealt with in some detail.

The implementation of these measures puts you in a much better position to carry out your design responsibilities concerning awareness, commitment and consistency of approach from project to project.

In **Chapter 3** we discuss many of the detailed aspects of the subject, including but not limited to those shown in the index to this book.

Chapter 4 looks at how you can integrate your health and safety activities into your design management system, then apply them to the appropriate degree as dictated by the scope of work and conditions applicable to a particular project.

Throughout this book a number of helpful techniques are described and examples provided. It concludes, in **Chapter 5**, with a review of the journey made.

Brian Thorpe, C.Eng, MIMechE, M.IEE, MIE.D, FIOA

Brian has had extensive experience at senior management level in engineering design, project management, sales and marketing, manufacturing and consultancy including more than 25 years in quality management.

He has worked within a wide range of British industries, including aerospace, nuclear, food, textiles and waste management.

He is an author and presenter of numerous training courses relating to a wide range of quality management subjects, including many relating to the construction industry, and has been a regular speaker on quality matters both within the UK and abroad. He has held the position of Quality Manager with several UK organizations and has been responsible for the application of quality management on some of the largest high-technology projects in Europe.

For six years he was head of one of the country's leading quality management consultancies, which was itself based within a construction engineering organization. Brian has helped numerous organizations of all types and sizes to develop 'right for them' quality management systems and gain third-party certification. For a further 12 years he was a registered Lead Assessor of quality management systems.

In addition to this book, Brian is co-author of *Quality Management in Construction* (Gower, 2004) and its two earlier editions, and is also the author of *Addressing CDM Regulations through Quality Management Systems* (Gower, 1999). He has also written many published articles on different aspects of quality management.

Now retired from full-time work activities, Brian continues to provide support to others, purely on a 'request for help' basis, where he feels qualified to do so.

You've got the book. To further enhance your knowledge and understanding of this very important subject we've developed this course.....

IOSH Health & Safety in Construction Design

The training is intended for those who take the legal role of designers and require the basic health and safety skills and knowledge that will enable them to fulfil their duties under CDM and other relevant legislation.

The course is workshop based and will draw on the experience of the delegates and tutor. Legislative requirements will be based on case studies of incidents and HSE prosecutions; the group will be encouraged to work together to develop innovative solutions to the problems posed.

Training objectives:

These are to ensure that at the end of the course delegates will be able to:

- Define hazards and risks and describe the legal requirements for design risk assessment.

- Identify construction hazards and high risk areas.

- Demonstrate a practical understanding of risk assessment techniques and requirements for recording the results.

- Apply the appropriate control measures for elimination or reduction of risks.

- Outline the main duties of designers and other duty-holders under the Construction (Design and Management) Regulations 1994.

- Outline the main provisions of the Health and Safety at Work etc. Act 1974 and the Management of Health and Safety at Work Regulations 1999 as applied to Designers.

- Outline other relevant health and safety legislation, codes of practice, guidance notes and information sources such as RIBA, CIC, ICE, IStructE, CIBSE, MaPS, the Health and Safety Executive etc.

- Establish procedures for the recording of required information.

- Establish an appropriate resource facility.

- A great opportunity to network and benchmark.

To book or for further information call

Tel: 08457 336666

National Construction College
Health, Safety & Supervisory

Chapter **1**

Regulatory Requirements

Background

There are a number of key publications which collectively embrace the main health and safety requirements applicable to the construction industry.

The following are just *some* of the main ones:

- Health and Safety at Work etc. Act 1974 (HASAWA)
- The Management of Health and Safety at Work Regulations 1999 (MHSWR)
- Control of Substances Hazardous to Health 1994 (COSHH)
- Manual Handling Operations Regulations 1992
- Construction (Health, Safety and Welfare) Regulations 1996 (CHSWR)
- Construction (Design and Management) Regulations 1994 (CDM).

There are others.

Although our prime interest in this book concerns the designer's role in helping to improve safety for people involved in, or affected by, construction and post-construction activities, in this chapter whilst not losing sight of that objective, we adopt a slightly different approach to it.

To do this we begin by taking a very brief look at different types of health and safety-related documents and the way in which they are viewed in law – that is, in Acts, Regulations, ACOPs, bulletins and so on.

Next we look in some detail at the above publications, considering in particular their main implications for on-site contractors, the objective being to try to identify key areas of responsibility where on-site (or post-construction) health and safety requirements should *also* form part of the designer's awareness and considerations – which basically constitute an appreciation of other's problems.

Here, we are not trying to produce design solutions to every potential risk problem (as a designer, you are not required to do that), but are trying to identify things which should be considered for inclusion in any design database, so that you can consider their applicability on a project-by-project basis.

Information is only of real value when people receive and understand it.

Arguably, much, if not all, the information you seek to produce already exists in the form of bulletins and the like. This is not totally true but, even if it were so, does everyone receive such information? And, if they do, what happens to it?

As we have already said, what really matters is not just the written word, but people's awareness and understanding of it and how it is used. For example:

- Is this information circulated to every person who needs to know?
- Is it the subject of periodic staff awareness/update training?
- Is relevant information extracted and incorporated into helpful health and safety-related databases, the contents and applicability of which people (such as designers) are required to consider, by virtue of their working procedures/instructions?

Or:

- Is the information merely incorporated into a technical library and a circular issued periodically to staff to advise that it is now available?

One of the most consistent comments arising out of HSE surveys has been about a perceived limited awareness on the part of designers concerning the obligations and problems of those with whom they work.

Let us now try to redress this situation, by looking on those you work with and who receive and rely on your output in order to do their job safely and right first time, as your internal (that is, within-project) customers. Let us determine to the best of our ability what *they* need, with regard to health and safety information, and in what form, then do our best to provide it.

Under the CDM regulations there is a vehicle for collating and communicating such information. It is called the health and safety plan.

Levels of importance of various health and safety-related publications

The Health and Safety at Work etc. Act (HASAWA)

In terms of legal standing is the Health and Safety at Work etc. Act 1974 takes precedence. As an Act of Parliament it comes under criminal law and, as such, must be complied with.

The Act is written in general terms, defining principles for good health and safety without prescribing or defining what specific measures should be taken to achieve them. Typically, the Act refers to the need for employers to do such things as:

- ensure in so far as is reasonably practicable, the health, safety and welfare of their employees whilst at work
- ensure a safe place of work and access to and egress from it
- provide for the safe storage, use and transport of materials
- provide protection from unnecessary risk or injury
- have a written (and maintained up-to-date) safety policy along with details of its implementation responsibilities and arrangements

- provide necessary training, information and supervision concerning health and safety matters.

There is a duty on employees to take some responsibility for their own health and safety as well as that of others who may be affected.

These requirements apply to every employer and workplace. Obviously the nature of application will vary depending on the prevailing circumstances. For example, within a city centre office, safety of access and egress may hinge around things such as functional maintained lifts, emergency stairways, clear exits and entrances, emergency notices for staff, well-defined escape routes, staff training, emergency drills, nominated safety officers and so on.

On a construction site, the same requirements will almost certainly have to take into consideration such things as separate traffic and personnel routes, avoiding work areas where risks exist due to the activities being carried out (for example, the risk of something falling on to someone, the use of barriers, no-entry zones and so on).

Although many sites may appear similar, each must be treated as unique and its specific health and safety requirements determined by techniques such as risk assessments carried out by suitably experienced people in advance of the start of work programmes. These must then be incorporated into a health and safety plan which becomes a basis for subsequent health and safety measures, training, coordination and so on.

This requirement is further covered in the discussion of the CDM regulations later in the chapter.

Regulations

Because the HASAWA 1974 is written in only general terms, something more is needed to amplify its interpretation in relation to different types of organization/industrial sectors and particular types of hazards relating to these. This is what the regulations aim to do. As far as the construction industry is concerned, those identified at the beginning of the chapter are some of the more significant.

The standing of such regulations is that they must be considered as compulsory and, as such, obeyed.

Codes of practice

These are prepared for the guidance of those who are required to comply with prevailing legislation as laid down within the above. Although they are not in themselves mandatory, they provide very sound qualified interpretations and advice concerning the carrying out of subject activities and, as such, will be of significant standing in the case of any legal proceedings.

Guidance notes

Such notes are frequently published by the relevant enforcement agencies (for example, the HSE) and provide useful professional guidance concerning the safe implementation of various subject activities. Compliance with such notes is neither mandatory, nor do they have recognized standing in legal terms. Nevertheless, they are worthy of serious consideration insofar as they are normally based on a wealth of experience and much feedback, and are also prepared by well-qualified people with the interests of others at heart.

A closer look at a selection of the key regulations

We will look at the following regulations from the perspective of their principal implications for contractors working on site and consider how prior thought and action by you, as designer, might help them meet their responsibilities and obligations.

The Management of Health and Safety at Work Regulations 1999 (MHSWR)

The key requirement is as follows:

Employers are required to make a suitable and sufficient assessment of risks to the health and safety of their employees while at work. This responsibility also extends to self-employed persons in respect of their own activities.

As a statement of requirement the above is very general, insofar that it begs many questions such as the following:

- What represents a hazard?
- What risks does it present?
- Are the risks significant or of a minor nature?
- How serious could be the consequences?
- What options are available for:
  - eliminating the hazard/risk entirely?
  - countering/reducing same?
- Which of these are reasonably practical?

Answering these questions is likely to be subjective in that one person's interpretation of risks and remedial options may differ from that of someone else.

This is where the availability of a comprehensive database of types of hazard, associated risks and possible solutions can be of value and help to harmonize thinking. Insofar as it applies to construction and post-construction stage activities, this database should not be significantly different in respect of either the contractor or the designer.

Both parties should appreciate, for example, that working at height is a hazard, what risk it represents, what the potential root causes of those risks are and what measures have been successfully used in the past or are available to counter them. With this depth of information, it is hoped that the designer will, as far as is reasonably practicable, 'design out' such hazards or risks, or otherwise 'design in' provisions to reduce or counter them.

The subjects of risk assessments and databases are dealt with in more detail in Chapter 3.

In circumstances where an employer has five or more employees, any significant assessment finding must be recorded, along with details of any group of employees deemed to be particularly at risk, the nature of the hazard, scope of the risk(s), any control measures in place and so forth.

Thus if a pre-construction assessment has been carried out by a designer and there remains the presence of a significant known hazard or risk (which could not be designed out), it is incumbent upon the designer to provide the fullest information possible, in order that the relevant details can be incorporated in the construction health and safety plan along with any measures subsequently taken as a result of the contractor's own risk assessment of the situation. Most of the requirements of the MHSWR are of a stand-alone nature – for example:

If you can't design out a risk, you need to design in ways to counter it.

- **Health and safety arrangements**
  There must be effective arrangements concerning the planning, organizing, controlling, monitoring and reviewing of systems and procedures pertaining to health and safety at work. Such arrangements should also be an integral part of the organization's management system.

  This is exactly what is advocated in Chapter 2 of this book and dealt with in some detail in Chapter 4 with particular emphasis on the design function. As far as communication between the designer and the contractor goes, part of the management system arrangements of both parties should go beyond internal matters and include provision for health and safety liaison as well as the timely passing and receipt of information (including provisions for feedback) as indicated in the final section of Chapter 2.

- **Health surveillance**
  This requirement to monitor people's health within their working environment is obviously likely to be more relevant to a site environment than that of a design office. Its purpose is to ensure that safe working conditions prevail and that people exposed to

Some health risks emerge over time from prolonged exposure to toxins, chemicals or unhealthy conditions.

health risks by virtue of their contact with, or proximity to, known dangers are subject to ongoing checks at regular pre-prescribed frequencies to determine that their health is not being affected as a result of temporary or continuing exposure to the risk source.

On a construction site typical risk situations may include:

- the handling of toxic substances
- chemical reactions resulting from the combining or working of certain materials
- fumes arising from a heat application process
- exposure to X-rays eminating from a radioactive source, such as that used for examining pipe welds
- risk of gas build-ups in confined spaces.

The list of potential hazards which can give risk to short- or longer-term health problems is extensive. Many may not always be apparent to people unless they have specialist knowledge. Everyone has their limitations. What can you, as a designer, do to help? There is a significant onus on the designer to:

- endeavour to gain an awareness of such potential health-threatening sources
- maintain a *dynamic* information base concerning substances, processes and so forth, which can give rise to such health risks and *refer to it*
- specify alternative materials/processes wherever possible
- consult with others (that is, seek expert advice) if in any doubt and feed back what is learned into the information base for the benefit of others
- provide maximum information, including suppliers' literature and recommendations
- incorporate appropriate warnings within the design information
- advise, if possible, on available health-monitoring techniques
- discuss with the planning supervisor and principal contractor, to help the latter ensure that appropriate surveillance arrangements are in place.

• **Health and safety assistance**
It is up to every employer to appoint a person (or persons) to assist in meeting the requirements of health and safety legislation.

• **Procedures for serious and imminent danger**
During the construction phase it is expected that suitable arrangements (for example, evacuation procedures) will have been established; that people will be familiar with their requirements; that the principal contractor will have ensured this; and that arrangements will have been coordinated as necessary. None of this is seen as involving the designer.

Beyond the construction phase, however, depending on the nature of the structure and its purpose, the designer, although not necessarily directly involved in determining procedures, should have considered potential underlying situations, insofar as they were foreseeable, and made provisions within the design (such as emergency stairs, fire escapes, sprinkler systems and alarms) in respect of possible dangerous situations occurring.

- **Information for employees**
  It is incumbent upon employers to provide employees with comprehensible and relevant information concerning any identified risks to their health and safety, any preventive and protective measures, and any emergency procedures. Beyond the provision of information concerning risks associated with normal duties a contractor can only meet this responsibility if they are aware of any other risks, preventive/protective measures and emergency procedures, which affect them.

  Under the CDM regulations (which are discussed shortly) such information should be embodied within the construction health and safety plan. An awareness of the contractors' need for such information should be an encouragement to provide the fullest information possible about known residual risks (that is, those that have not been fully designed out) and to liaise closely with the planning supervisor and principal contractor in order to help establish measures to combat these risks.

  Other requirements under the MHSWR, namely:
  - cooperation and coordination between employers
  - working on other premises
  - capability and training
  - employees' duties
  - temporary workers

  are not seen as requiring any special designer input.

Control of Substances Hazardous to Health (COSHH)

The number of substances which can be hazardous to health as a result of exposure to them in the course of work activities is considerable, especially when you consider that the regulations apply to both natural and artificial substances (and their preparation) in various states – namely, solid, liquid, gaseous or vapour. Micro-organisms also fall under COSHH.

Lead and asbestos are not included as they are subject to their own separate regulations.

This is an area where front-end activities, such as design, can have a major influence on the health of those downstream, who are involved in the actual build/construction processes. The following steps should be taken:

- Avoid, where possible, the specification of substances which are hazardous in their own right or give rise to health hazards during their processing.

• Where the above is not possible, provide the planning supervisor with the fullest information concerning the hazard so that the contractors can be made aware and put in place appropriate health protection measures.

Don't try to hold all this information in your head; build up a database of information on potential hazards.

Your problem, as the designer, is one of 'awareness': when there are so many diverse types and sources of such hazards, how could you be expected to be aware of them all? It is just not possible. In theory, you would need to be aware of hazards which may exist on a site as a result of ground contamination, possible gaseous build-ups and emissions (such as methane from a recovered refuse disposal site), hazards which may be inherent in the dismantling of existing structures, materials or processes which can be safely used in construction, other materials which present known hazards requiring specific control measures, as well as others which should not be considered under any circumstances. To have any hope of making a real inroad into the hazardous substances situation, it is essential to have access to a really good design database of information.

Where, however, should such data come from?

The following are seen as potential sources:

• Any existing health and safety file relating to the intended construction site, which should contain information about any known hazards.
• A comprehensive checklist, indicating hazard/risk-related sources about which information needs to be available or should be obtained, in order that such potential hazards can be discounted or alternatively included as appropriate in any assessment of risks.
    The key purpose of such a list is twofold:
  – To provide an *aide-mémoire* for the benefit of designers regardless of their experience.
  – To introduce an awareness of hazard potentials which may otherwise be overlooked due to the limited/variable knowledge of individual designers.
• An extensive (ideally computer-accessed) database, as previously mentioned, built up from the mass of information already in the

Nature of hazard	Associated risk(s)	Indicated action/measures	Reference information

public domain and presented in a user-friendly manner such as the example below.
Obviously such a database could be extensive, with hazardous substances being just one section of it.

Some clients may produce lists of substances that may not be used on projects being carried out on their behalf. Such lists must be adhered to and ideally captured on the aforementioned database to promote awareness and consideration on future projects.

Manual Handling Operations Regulations 1992

These regulations came into force on 1 January 1993. Their aim was to prevent injuries resulting from manual handling operations. In general terms, manual handling concerns those situations where human force is required to move or hold a load whether applied directly (as when lifting a weight by direct contract) or indirectly (as when using a pulley or applying leverage).

The regulations require that, so far as is reasonably practical, the need for hazardous manual handling be avoided.

Where such handling is unavoidable then the situation must be subject to an assessment of associated risks (if not already included in, and clearly cross-referenced to, another assessment).

As with hazardous substances, you would need to be knowledgeable on a wide range of issues and also have an understanding of the human anatomy, average lifting or force-applying capabilities, potential sources of overload, bodily stress and so on, in order to effectively carry out a detailed risk assessment of each manual handling situation.

Your main aim must be to avoid manual handling as much as possible, deploy suitable lifting/handling devices and also provide as much information as possible concerning loads, weight distributions, centres of gravity, lifting points (and provisions for same). This will essentially involve joint discussions with the planning supervisor and principal contractor for the reasons given in Chapter 3.

Construction (Health, Safety and Welfare) Regulations 1996 (CHSWR)

These regulations are extensive in their coverage and apply to all construction work irrespective of its duration or the number of workers involved.

Construction work is described as:

Any building, civil engineering or engineering construction work, including:
construction, alteration, conversion, fitting out, commissioning, decommissioning, renovation, repair, maintenance

(including high pressure cleaning) redecoration, demolition or dismantling.

Also covered is site clearance at the start and finish of the construction work, the assembly and dismantling of prefabricated units and the installation, maintenance and removal of services associated with a structure (for example, mechanical, electrical, gas and so on).

A structure is described as:

Any buildings, steel or reinforced concrete structure (not being a building), railway lines or sidings, tramways, docks and harbours, inland navigation channels, tunnels, shafts, bridges, viaducts, waterworks, reservoirs, pipes and pipelines, cables, aquaducts, sewers, sewage works, gasholders, roads, airfields, sea defence works, river works, drainage works, earthworks, lagoons, dams, walls, caissons, masts, towels, pylons, under-ground tanks, earth retaining structures, structures designed to preserve or alter any natural feature or any structure similar to the above, formwork, falsework, scaffolds or other supporting or access structures, any fixed plant where its installing, commissioning, decommissioning or dismantling can involve the risk of a person falling more than 2 metres.

These descriptions as to what constitutes construction work and a structure make the extent of the CHSWR's application very clear. Indeed, it is difficult to imagine a construction project which does not involve a combination of several of the factors described.

The list below gives a brief interpretation of the main require-ments of the CHSWR and will hopefully help provide the designer with a greater awareness of some of the site risk situations and obligations faced by contractors and others (for example, self-employed workers) working at the 'sharp end' of the construction process. Many of the health and safety requirements described will automatically fall under functional design requirements – for example, the inclusion, where necessary, of fire detection and sprinkler systems. Others, however, are things which are felt to have a direct bearing upon health and safety during construction and post-construction and where the designer could have a positive influence on improving health and safety. These points have been indicated by an asterisk (*).

Communication and cooperation are vital if regulatory requirements are to be understood and met.

- Compliance with the regulations is mandatory upon employers of employees carrying out construction work and also on self-employed persons and any other person (who is not an employee) carrying out such work (Regulation 4).

- All persons are to cooperate with other persons to ensure that the regulations are complied with (Regulation 4).

- There is a general requirement to ensure a safe place of work with safe access, egress and sufficient working space with regard to the nature of the work being carried out (Regulation 5).*

- So far as is reasonably practicable, steps must be taken to prevent falls from places of work or the means of access to, or egress from such places, by the provision of suitable and maintained methods, such as guard rails, toe boards, barriers, working platforms and so on. Where the above is not practicable, suitable secured personal suspension equipment must be provided or, if none of the foregoing is practicable, then suitable fall arrest measures must be adopted (Regulation 6).*

- Suitable steps must be taken to prevent persons falling through fragile materials (that is, any material likely to break under the weight of a person (or other load) upon it (Regulation 7).*

- In respect of potential falls of two or more metres, warning signs are required, indicating approaches to fragile materials and where it is safe for persons to pass across or near to, or to work or near to such materials (Regulation 7).*

- Steps must be taken to prevent objects or materials falling on to any person (Regulation 8).

- Throwing or tipping objects or materials from heights where a danger of injury to others may occur, or the storage of such objects and materials in a manner which has a potential for danger in event of their movement or collapse, is not permissible (Regulation 8).

- Steps must be taken to prevent the accidental collapse of new or existing structures or any part thereof, due to their becoming weak or unstable as a result of construction activities. In other words, structures must not be overloaded. Temporary support structures and similar must be created and dismantled under competent supervision (Regulation 9).*

- Demolition/dismantling must be carried out under the supervision of a competent person (Regulation 10).

- Explosive charges may only be fired following the necessary steps to ensure that nobody is exposed to risk or injury (Regulation 11).

- Steps must be taken to prevent dangers arising from the collapse of new or existing excavations.
  Also, suitable and sufficient steps must be taken to prevent persons being buried or trapped as a result of the fall or dislodgement of excavation material – for example, side/roof supports. These must only be installed, altered or dismantled under the supervision of a competent person.*

- Risks of persons or vehicles falling into the excavation and consequent risk to those working in the excavation must be avoided, as well as those of instability to the excavation itself due to the proximity of persons or vehicles operating in close proximity.*

- Dangers must be taken into account with respect to underground cables and services (Regulation 12).*

- Cofferdams and caissons must be of suitable design and construction for their intended purpose and suitably maintained.*

- Erection, installing, alteration or dismantling must be done under competent supervision (Regulation 13).

- Steps must be taken to prevent people falling into water or other liquid as far as is reasonably practicable.*

- Suitable and properly maintained rescue equipment must be available immediately.

- Any means of transport across water must be suitably constructed, maintained and under the control of a competent person (Regulation 14).*

- The site must be organized so as to ensure the safe movement of both vehicles and pedestrians, with routes suitable and sufficient for the users.*

- Doors or gates leading on to vehicular routes should enable good visibility of approaching traffic from a safe viewpoint.*

- Vehicle routes must be clear of obstructions.

- Routes must carry suitable signs for reasons of health and safety when necessary (Regulation 15).

- Doors and gates which could give rise to danger must be fitted with suitable safeguards (Regulation 16).*

- Suitable steps must be taken to prevent unintentional vehicle movement.

- Arrangements must be made for giving warnings of any possible dangerous vehicle movements (such as reversing or tipping).

- Care must be taken to prevent vehicles tipping over (into excavations, for example).

- Safe operation, including prohibiting of riding in/on vehicles during loading/unloading of loose materials, must be ensured (Regulation 17).

- Suitable and sufficient arrangements must be made to prevent risks of injury as a result of fire, explosion, flooding or asphyxiation (Regulation 18).*

- Emergency routes and exits must be provided, kept clear and indicated.

- Emergency procedures must be in place, including those for evacuation where necessary.

- Persons must be made aware of the procedures, which must be tested (Regulations 19 and 20).

- Suitable and sufficient fire detection, alarms and fighting equipments must be located properly.*

- All the above must be maintained and tested at regular intervals.

- All persons on site must receive instruction in the use of relevant fire-fighting equipment.

- Work involving fire risk must only be carried out by persons who have been trained in necessary risk prevention (Regulation 21).

- Compliance with welfare requirements must be met – for example:
  - suitable sanitary and washing facilities
  - drinking water supply
  - rest facilities
  - changes and storage of clothing (Regulation 22).

- There must be adequate provision of fresh/purified air at every workplace.

- Equipment providing the above must be fitted with a device/devices to give audible or visual warning of its failure.

- Internal workplace temperatures must be maintained at reasonable levels.

- External workplaces must be arranged so as to provide protection against adverse weather conditions, including, when necessary, the provision of protective clothing (Regulations 23 and 24).

- Suitable and sufficient lighting must be available, including the provision of secondary lighting in the event of there being a risk to health and safety if the primary lighting source were to fail (Regulation 25).

- The construction site must be kept in good order.

- Boundaries must be marked with suitable signs (Regulation 26).

- Plant and equipment must be safe, not present risks to health, and be well constructed and maintained (Regulation 27).

- Construction activities requiring training, technical knowledge or experience in order to reduce risks to health and safety must be carried out by persons appropriately qualified, or otherwise supervised by such a person (Regulation 28).

- Before the commencement of work at height, on excavations, cofferdams and caissons begins, the place of work must be inspected (with further inspections at designated frequencies) by a competent person, who must be satisfied that it is safe to carry out work.

- Following inspections, a competent person must raise a written report (Regulations 29 and 30).

Construction (Design and Management) Regulations 1994 (CDM)

These regulations came into effect on 31 March 1995 and identify the key players within a construction project and the discrete responsibilities associated with each. They cover a broad range of activities relating to the carrying out of building, civil engineering or engineering construction work including the following:

- The construction, alteration, conversion, fitting out, commissioning, renovation, repair, upkeep maintenance (including cleaning using water or abrasive at high pressure or substances classified as corrosive or toxic), decommissioning, demolition or dismantling of a structure.

Additionally applicable are:

- The *preparation* for an intended structure – for example, site clearance, exploration, investigation and excavation (excluding site survey) and laying or installing the foundations of the structure.

The CDM regulations cover preparing the site, assembling the elements and removal or demolition.

- The *assembly* of prefabricated elements to form a structure or the disassembly of such elements which, immediately prior to disassembly, formed a structure.
  (**Note:** If the above were being carried out in an 'off-site' workshop, then the regulations, except with regard to designers, would not be deemed applicable.)

- The *removal* of a structure or part-structure or of any product or waste resulting from demolition or dismantling of a structure or from the disassembly of prefabricated elements which, immediately before such disassembly, formed a structure.

- The installation, commissioning, maintenance, repair or removal of mechanical, electrical, gas, compressed air, hydraulic, telecommunications, computer or similar services which are normally fixed to a structure.
  (**Note:** The exploration for or extraction of, mineral resources or activities preparatory thereto carried out at a place where such exploration or extraction is carried out, are not included.)

What is a structure?

Again, we are faced with a very broad list of inclusions, of which the following is indicative (not definitive): any building, reinforced concrete structure, rail or tramway lines, sidings, docks, bridges, viaducts, aquaducts, pipelines, reservoirs, sewage works, gas holders, airfields, sea defences, dams, caissons, masts, pylons and so on.

Also included is formwork, falsework, structures that provide means of access, and fixed plant where, during installation, commissioning, decommissioning or dismantling, there is a danger of someone falling more than two metres.

Applicability

In *general terms*, the regulations apply if:

- the construction phase lasts more than 30 days
- the construction phase involves in excess of 500 person days
- construction work entails the demolition or dismantling of a structure
- five or more persons are at work at any one time.

For any of the above, formal notification to the HSE is required.

The regulations *do not* apply to construction work on domestic premises unless they are used for business purposes – although the duties on the designer and to notify the work (if appropriate) remain – where a local authority is the enforcing authority for health and safety and the work is of a minor nature.

Who are the key players?

The key players are the client, planning supervisor, principal contractor, contractors and designers. However, the number of permutations with regard to these rules which can result is a little confusing.

For example:

- It is permissible for the client to delegate the responsibilities imposed under the regulations in total or in part to another party – for example, an agent. In the case of such an appointment, a formal declaration of the same has to be made to the HSE in accordance with Regulation 4. The HSE in turn is required to formally acknowledge this declaration.

- Subject to adequacy of competence and resources, the client may appoint the same person as both planning supervisor and principal contractor, or the client can carry out one of the above roles, or carry out the role of the designer directly.

- The designer may be a stand-alone function, or part of a contractor organization, which may have been asked to fulfil a design and build role.

- The principal contractor may not always be the main contractor on a project, although logically this would seem preferable.

You are advised to familiarize yourself more fully with these roles by studying the regulations and associated ACOP. For now, rather than becoming bogged down with permutations and possibilities, we will move ahead and look at the duties vested in each key role on a typical notifiable project.

The client's duties

- Appoint a planning supervisor and a principal contractor (subject to being satisfied as to their competence and ability to commit the necessary resources).
- Provide information about the construction site to the planning supervisor.
- Ensure that an adequate health and safety plan has been prepared by the principal contractor prior to the start of construction work.
- Ensure that the health and safety file is kept available for reference by those needing to access the information contained within it.
- Transfer the health and safety file to any new owner of the property or structure and ensure that the recipient is fully aquainted with its purpose.

These duties can be elaborated on as follows.

An effective planning supervisor will be someone who has experience of both design and contractor functions.

- **Appointing the planning supervisor**
  The planning supervisor is a new position and is effectively an addition to the project team, albeit accountable directly to the client. Because the appointee is effectively forming a health and safety-related link with both designer(s) and principal contractor the chosen person should ideally have good experience in both the design and contractor functions.

  This experience, plus technical, coordinating and communicating skills, will collectively be significant factors in assessing competence.

  The other major consideration is their ability to resource the necessary duties to be undertaken. Resources may include such things as computer capabilities and adequate time allocation.

  This particular appointment should be made as early as possible, once site information has been compiled.

- **Appointing the principal contractor**
  This appointee must be a contractor by profession. Their role is essentially that of providing health and safety guidance and

coordinating service throughout the whole of the construction phase.

Again, an absolute requirement for the job is an ability to demonstrate the competence and commitment of the necessary resources to fulfil what can be a very demanding role.

- **Providing site information to the planning supervisor**

This is a vital early function which the client must consider very seriously and make every effort to fulfil as effectively as possible. Close liaison with the planning supervisor (who should also be liaising with any existing designer) is important because the planning supervisor will need to make this site information available to the designer (if not already known to them), thus providing details of potential hazards and guiding the designer in their assessment of risks and the elimination of them.

The outcome of this planning supervisor/design liaison will help the former prepare an initial pre-tender health and safety plan for the guidance of the principal contractor and others.

The following is a list of typical (but not definitive) information that you may provide to the planning supervisor (see CDM Regulation 11(1):

- address of premises
- premises description, including site plans, maps and so on
- details of any existing structures or plant
- planning history (if known)
- any current planning conditions concerning the premises
- names or identities of any tenants or users of the premises
- copies of any surveyor, feasibility or other reports prepared on behalf of the client
- any past industrial useage history
- details of any relevant communications with the HSE regarding the premises
- notices of any litigations or prosecutions pertaining to the premises or others adjacent to it
- client future requirements
- client budget
- any existing health and safety file.

- **Ensuring that an adequate health and safety plan has been prepared by the principal contractor prior to the start of construction work**

It should be noted here that the word is 'ensure', not 'approve'. It is likely that a client may solicit the advice of the planning supervisor on this matter.

It is also possible that separate health and safety plans may be needed for different structures. On a large, long-duration project, some work may reach its construction stage while later work is still in design. Because work needs to continue in accordance with planned programmes, separate plans may be necessary for the successive stages.

- **Managing the health and safety file**

  The health and safety file is an important compilation of project-related health and safety information. It forms a valuable record which is available for reference beyond the project completion date (for example, for providing essential design input or health and safety plan information for any further development work) and for the total or partial handover to any person (such as a future site or facility owner) with a right to receive such information.

  It is incumbent upon the client to maintain this file and ensure its availability for transfer to others if it becomes necessary. It is also the client's duty to make any new recipient aware of its purpose.

  Some typical inclusions in a health and safety file are listed below:

  - historic site data
  - pre- and post-construction phase site survey information
  - photographic record of essential site elements
  - statements of design philosophy, calculations and applicable design standards
  - drawings and plans (including drawings prepared for tender purposes) used throughout the construction process
  - drawings and plans of the complete structure
  - maintenance instructions
  - instructions on equipment handling or operation, plus relevant maintenance instructions
  - results of proofing or load tests
  - commissioning test results
  - materials used in the construction, with particular emphasis on any hazardous material(s) including data sheets prepared and supplied by suppliers
  - identification and specification of in-built safety features (for example, fire-fighting systems, fail-safe devices)
  - method statements produced by the principal contractor and/or other contractors.

With respect to the client's duties, two notable omissions concern the appointment of designers and contractors. Whereas the client has a direct responsibility to appoint, in the case of the planning supervisor and principal contractor, the responsibility with regard to the appointment of designers or contractors is one of being reasonably satisfied as to their competence to fulfil their obligations under the regulations, including the allocation of appropriate resources to do so.

Often the first appointment made by the client will be that of the designer – for example, an architectural practice – whose initial duty may be to develop alternative design proposals for consideration or

The health and safety file is an important reference for the facility's owner long after the building is complete.

develop a conceptual design up to the stage where a firm decision to proceed can be taken. In such cases it is likely to be the designer who decides that the project is notifiable and reminds the client of the duties to be met under the CDM regulations, including those of appointing a planning supervisor and providing the site information needed for risk assessments and preparation of a pre-tender health and safety plan.

The planning supervisor's duties

These can be summarized as follows:

- Notify the HSE in writing, giving specific information applicable to notifiable construction projects.
  (**Note:** The information in question is given in schedule 1 to the CDM regulations.)

- As soon as is reasonably practicable and before the start of construction work, obtain information concerning the site (see the previous section on the client's duties for the typical information required).

- Advise the client, as necessary, on the competence and ability to commit necessary resources, of other appointees, for example, contractor(s).

- Liaise closely with the designer(s) to ensure that their obligations concerning health and safety as required under Regulation 13 are being met.

- Prepare a pre-tender health and safety plan to provide information to the principal contractor (when appointed) concerning the site, known hazards and so on.

- Ensure that a health and safety file is prepared in respect of each structure included in the project (Regulation 14 (d)).
  (**Note:** The planning supervisor does not have to produce the file personally, but he has to ensure that the information it contains meets the requirements of Regulation 14(d)(i) and (ii).)

- Review, amend or add to the health and safety file as necessary, to ensure that it contains all the necessary information when passed on to the client.

The principal contractor's duties

The duties of the principal contractor (who must be a contractor by profession) take into account the meeting of the many requirements we have described earlier, when discussing a selection of the more relevant regulations and their applicability to site activities. They are as follows:

- First and most important, prepare the construction health and safety plan which defines or references all the health and safety arrangements for managing the work.

- Ensure that the plan is available for review by the client (or their nominated representative) prior to the commencement of any related construction work.

- Carry out risk assessments and review those carried out by other contractors to ensure that they adequately address risks to their own employees and others who may be affected.

- Where a common risk to several contractors is apparent, coordinate the preparation of a single risk assessment.

- Ensure that measures arising from risk assessments are incorporated in the health and safety plan.

- Be reasonably satisfied if or when appointing others, that they have the competence and resources to carry out their obligations under the regulations.

- Ensure that others are fully aware of any risks contained in the health and safety plan.

- Take reasonable steps to ensure that only authorized persons are permitted to enter any premises, or part thereof, where construction work is being carried out.

- Decide upon the adequacy of individual contractors' health and safety arrangements and arrange for their coordination if necessary. Incorporate them into the health and safety plan.

- Do the same for the safe use of mutually shared equipment.

- Do the same for emergency procedures.

- Ensure that arrangements are in place to allow others to discuss health and safety issues and offer advice. Also arrange for the coordination of the views of employees and their representatives.

- Establish and maintain a good health and safety records system, such that any necessary information needed for the health and safety file can be assembled.

Other duties concern ongoing liaisons with the planning supervisor, other contractors, the posting of site notices identifying the names of persons occupying key posts and so on.

From the foregoing it will be apparent that the principal contractor needs to be an organization with:

The principal contractor must be able to manage, communicate and coordinate health and safety issues effectively.

- good awareness of the responsibilities of the principal contractor
- project management experience

- good communication and coordination skills
- an ability to give guidance to, and encourage participation by, others concerning health and safety matters
- a dedication to health and safety
- the ability to select/appoint/recommend others.

The contractor's duties

These can briefly be described as follows:

- Have and maintain a sound health and safety policy.

- Seek information concerning the applicable requirements of the health and safety plan, the identities of the planning supervisor and principal contractor and ensure that their own employees or self-employed persons are provided with this information, before they are permitted to begin construction work.

- Carry out risk assessments and take all the measures necessary to eliminate or reduce such risks. (This is a requirement of the MHSWR.)

- Cooperate with the principal contractor so far as is necessary to enable both parties to comply with their duties under the relevant statutory provisions. This should include passing any information needed to comply with instructions received from the principal contractor or with any directions in the health and safety plan.

- Operate effective health surveillance.

- Ensure that employees and self-employed persons are able to discuss and offer advice concerning health and safety.

- Be reasonably satisfied regarding appointees' competence and resources to carry out their duties under the regulations.

The designer's duties

These are addressed in Regulation 13 which is applicable whenever the design work in question relates to construction work, regardless of whether or not the project is notifiable. This regulation is applicable to the design of construction products, which may not be assigned against any particular project – for example, made for stock and future sale.

In general terms, Regulation 13(1) requires the designer to take reasonable steps to ensure that the client for the project in question is aware of the duties of a client by virtue of the regulations, plus any practical guidance issued from time to time by the Health and Safety Commission with respect to the requirements of the regulations.

This is an important duty and it is recommended that it be done in a formal manner – for example, by letter.

Regulation 13(2)(a) requires that every designer ensure that any design they prepare, which they are aware will be used for the purposes of construction work, includes among the design considerations adequate regard to the need to:

- avoid foreseeable risks to the health and safety of any persons carrying out construction or cleaning work in or on the structure at any time or of any person who may be affected by the work of such a person
- combat at source risks to the health and safety of the above persons
- give priority to measures which protect all persons over measures which only protect each person carrying out such work.

Regulation 13(2)(b) requires that the designer ensure that the design includes adequate information about any aspect of the project or structure or materials which might affect the health and safety of any person as described above.

Regulation 13(2)(c) requires cooperation with the planning supervisor and also with any other designer who is preparing any design in connection with the same project so far as is necessary to enable each of them to comply with the requirements and prohibitions placed upon them under the relevant statutory provisions.

Summary

So far, you should have gained an appreciation of a number of the main regulations which have to be met by those working in construction. In doing so, you will also have gained a better feeling not only for some of the sources of hazards that exist, but also for aspects where consideration by the designer could be of significant help in easing the burden of compliance on the contractor.

In the case of the CDM 1994 regulations we have seen how discrete actions by, and planned cooperation between, the key players can help bring about improvements in relation to health and safety in construction.

Everyone needs to see themselves as an important link in the health and safety performance chain and realize that the improvements we seek will only be brought about by everyone playing their part and understanding what that part is.

The role and responsibilities of the health and safety inspector

Health and safety inspectors have the authority to enforce the Health and Safety at Work etc. Act and the regulations which underpin it.

Appointed by the HSE, the inspectors have extensive powers, as described below in summarized form.

- To enter premises at any reasonable time if there is suspicion of a dangerous situation.
- To seek the accompaniment of a police officer if necessary.
- To be accompanied by any other person authorized by the HSE along with any required equipment.
- To quarantine.
- To take measurements, photographs or samples of articles or substances.
- To carry out tests.
- To dismantle plant or equipment.
- To take possession of articles/substances so that they cannot be tampered with or to ensure availability for evidence in subsequent legal proceedings.
  (**Note:** An explanatory note of such actions is to be left by the inspector in event of such action.)
- To question individuals.
- To inspect or take copies of documentation.
- To demand facilities and assistance to enable the carrying out of their duties.
- To issue 'improvement' and 'prohibition' notices.
  (**Note:** Appeal against such notices can be made within a limited period (21 days) of such notices being raised. Failure to comply, however, can lead to a significant fine, or imprisonment may be imposed on conviction.)

Chapter 2

The Design Process

The multiple responsibilities of the designer

There is no stage in the construction process which is more important than that of design. It is at this stage that conceptual ideas are converted into constructable realities. It is at this stage, also, that a variety of considerations, each of which impact upon the others, need to be balanced simultaneously. Among the most important of these are:

Designing for safety needs to be an integral part of every stage in the wider design process.

- design for function (over a projected life period)
- design for construction
- design for safety
- design for maintainability
- design for cost
- environmental aspects
- aesthetics.

We cannot really design for any one of the above without taking into consideration the effects on all of the others. This means that designing for safety is not a stand-alone consideration, but an integral part of the wider design process. The more widely this is recognized and catered for within design planning, the safer will it become to construct and maintain structures and facilities.

It is at the design stage that:

- specified requirements are translated in construction configurations
- materials are selected
- safety factors are established
- levels of stress are determined
- fabrication/erection techniques are dictated
- safety considerations are applied
- regulations, legislation and codes of practice are complied with
- environmental issues are considered
- maintenance provisions are defined

amongst many other considerations.

Everything after the design stage is essentially concerned with meeting design requirements. If the basic design is flawed, the final result will be compromised, possibly with far-reaching, or even disastrous, consequences.

Perceived shortcomings in design in relation to health and safety in construction

Since the CDM regulations came into effect in March 1995, health and safety-related design activities in the construction industry have come under increased scrutiny.

Various professional bodies, including the HSE, have carried out audits and surveys which have raised concerns about the perceived lack of designers' awareness regarding their responsibilities under CDM Regulation 13 (see the Preface).

If we accept this situation and look for reasons why it should be so, we find that these can be numerous. Figure 2.1 represents a very simple 'cause and effect' diagram. It illustrates just some of the factors which, in isolation or combination, are seen as potential contributors to the reported findings. Let us, then, as designers, attempt to address the situation.

We have already listed some of the main considerations that the designer has to balance when producing a design. In terms of designing for safety, however, the designer's thinking will predominantly and understandably be directed towards ensuring functional safety throughout the lifetime of the subject construction.

The last thing any designer wants is a failure that could result in a disaster, leading to serious loss of life, loss of facility, huge costs, reputation damage and so on plus (in certain types of industry) a problem that could persist for years or even generations – for example, a nuclear disaster with widespread radiation effects.

The history of the construction industry is littered with spectacular disasters or structural failures covering a whole range of situations: from the collapse of almost all the buildings in townships caused by earthquakes, bridges moving, swaying and collapsing as a result of vibration frequencies imposed by winds or traffic, cooling towers shearing off, dam slippages due to unstable foundations and so on. In some cases, these failures were caused by factors which existed before design, but nevertheless manifested themselves in terms of inadequate design inputs.

The causes of major failures can be traced back to pre-construction activities.

Interestingly, surveys that have been carried out on all types of major disasters (not necessarily construction-related) have revealed that in the majority of cases (approximately 70 per cent), the root cause could be traced back to things which should have been done, but were not done, or were at least done inadequately, before any practical work had been carried out. Typical root causes were seen as:

- inadequate research and development (that is, going ahead on an assumed premise)
- poor/inadequate specifications
- poor planning
- lack of training.

It is interesting to note how essentially the same points appear in Figure 2.1.

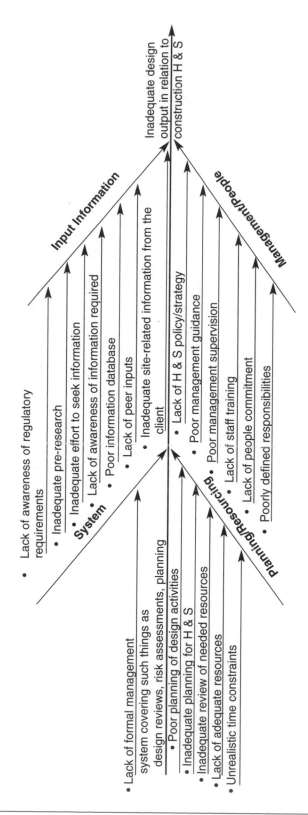

Figure 2.1 *Simple cause and effect diagram, showing some (not all) typical root causes of substandard design output*

As a further comparison with Figure 2.1, it is worthwhile comparing the methodology used by the HSE with regard to its own design/safety in construction audits. The approach used involved looking at factors that underpinned the actual design, namely:

a) policy, documentation and procedures (that is, basically the management system)

b) awareness of CDM requirements

c) training in CDM (and health and safety)
   (Both (b) and (c) come back to having sound management systems, with clear management responsibilities for establishing a health and safety policy, training, effectiveness monitoring and so on.)

d) feedback and how corporate knowledge is captured and disseminated
   (Again, this is part of any decent management system, which will include provisions for: collecting performance information, feedback, analysis, using the data for improvement, action initiation and effectiveness review.)

e) practical implementation
   (This concerns awareness training, application, recording effectiveness, management responsibility and so on.)

f) integration into design
   (Once again, this involves the management system and procedures with inclusions for health and safety requirements.)

The indicated solution: a health and safety culture

Wherever we look, the indicators all seem to point in one direction. If we are to really achieve an equivalent degree of design attention to safety during the construction and subsequent maintenance and cleaning stages of a project, as that given to functional safety, a number of things need to happen.

- You need to have a clear policy within your design organization, authorized by senior management, which commits to achieving health and safety during construction and recognizes and adheres to the CDM regulations and other applicable regulatory requirements.

- You need to have a strategy for implementing the policy through a well-presented (and monitored) management system, defining health and safety measures to be applied at every stage of the design process.

- All staff within the organization must be formally trained in the importance of health and safety to the business, themselves and other stakeholders (such as the public). Such training should include appropriate tests of understanding. In other words, your aim must be to establish a total health and safety culture.

- You need to create a management system which defines both organizational and functional responsibilities, and to develop procedures that include health and safety-related activities, alongside those for effectiveness review, performance monitoring, feedback and continuing improvement.

- You need to establish a comprehensive and dynamic health and safety information database, including properly indexed publications and other data, which is continuously developed in the light of feedback based on experience and peer inputs.

The following pages develop the above key requirements and offer suggestions on how to proceed.

Communicating health and safety messages involves checking that everyone has understood and remembers them.

Defining policy

You need to begin your journey right at the top with the organization's senior management and their responsibility for establishing a heightened awareness of health and safety in construction, maintenance and cleaning and the part that the designer can play in bringing this about.

If your organization already has a formal quality management system (as most do), it will almost certainly include as its introduction a 'policy' or 'mission' statement. Any such statement should be reviewed to see whether it needs to be more inclusive, making it clear that it is the organization's declared commitment to meet fully its obligations under the CDM (and other relevant) regulations, including due considerations for the health and safety of those who may be involved in the construction, maintenance and cleaning of the structures it designs, and also of others who may be affected, such as the public at large.

If no such policy statement exists, one should be introduced. It should be signed by the chief executive of the organization or the executive with delegated responsibility for health and safety management and should then be brought to the attention of all staff by:

- including it in appropriate manuals
- displaying it on noticeboards
- incorporating it into in-house training.

Health and safety information needs to be visibly brought to people's attention.

An attainment strategy

Your policy statement is merely one of intent. By itself it is not going to make things happen. If it is going to be met, you will need to have a strategy, backed up by an action plan in which separate responsibilities, activities and methods are defined.

What, then, will be your strategy? How will you ensure that your policy is carried out?

The following steps are proposed.

Step 1: Implement a training programme

Your aim will be to achieve a common, organization-wide understanding of the importance of health and safety and to establish, if you can, a consistent (best) approach to meeting your responsibilities, rather than a subjective approach. That is, the training should aim to promote a wider awareness of safety issues (that is, beyond functional safety) as well as an increased consideration for the risks to, and needs of, others downstream in the construction process within the bounds of practicality.

Ideally, the training programme should be organized by a senior executive and cover the following:

- why health and safety is seen to be of such importance to the organization (see the Preface)
- the CDM regulations, key parties, responsibilities and interfaces, with emphasis on CDM Regulation 13 (see Chapter 1)
- other key applicable regulations (see Chapter 1)
- the findings of surveys and audits carried out by the HSE and others in relation to design organizations (see the Preface and Chapter 1)
- what is expected of the designer and what isn't (see Chapter 1)
- the importance of having and using a well-maintained health and safety information database (see Chapter 3)
- how it is intended to establish and/or develop your own database (see the next section and Chapter 3)
- how it is intended to plan your health and safety responsibilities into your management system, so that the necessary actions can be taken, become part of the standard organization approach and be subject to ongoing review (see Chapter 4).

Step 2: Establish a health and safety database

Here your aim is to establish as comprehensive a database as possible.

The database should hold different types of information and be dynamic in terms of incorporating new information concerning hazards, risks, and potential solutions as soon as they become apparent.

The objective is to enable every designer, regardless of their length of experience, to have access to information (with guidance on how to use it). This will promote consideration of:

- what hazards do or may exist and therefore need to be taken into account
- what, if any, proven options for solution are known and could be considered.

Ideally, the information will be presented in a way which relates it to specific stages of the design process and beyond. For example:

The information in your database needs to be easily accessible to everyone and relevant to different stages in the design process.

- information to be established or considered, about the intended construction site – in other words, information which will constitute a major part of the initial design input

- information relating to construction materials and the hazards represented by the use and processing of these materials

- information relating to fabrication and assembly processes

- information pertaining to dangers inherent in certain types of maintenance or cleaning activities

- information to promote lateral thinking with respect to hazards generated as a result of the construction (however successfully completed) for others either near to or remote from the site, long after the completion and commissioning of the facility.

  Figures 2.2 and 2.3 show typical examples of how such problems may be caused.

The information detailed above can be obtained from sources such as: Croner, health and safety publications, HSE bulletins, information produced by the professional institutions, industry, and so on plus inputs from experienced colleagues, multi-organizational reviews and discipline reviews, as well as feedback from real-life experiences.

Obviously, the database information should be easily accessible and user-friendly. It also needs to be controlled by a specific individual, responsible for ensuring its updating and general management.

Note: Most organizations will have databases relevant to their range of responsibilities. In this particular instance we are talking about a typical architectural practice or design office.

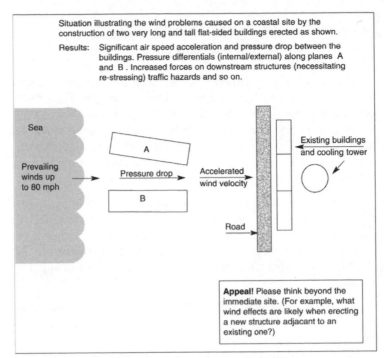

Figure 2.2   *Typical example of how a construction can cause post-construction problems for others*

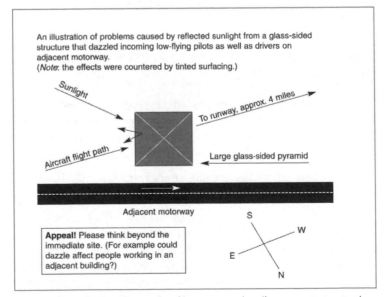

Figure 2.3   *Further example of how a construction can cause post-construction problems for others*

Step 3: Plan and resource health and safety requirements into the design programmes

If health and safety-related activities are to be done, and seen to have been done, it is no use leaving them to chance. They need to be visibly planned into design programmes and allocated resources of time, people and money. The CDM regulations emphasize the provision of necessary resources to implement health and safety requirements. There is no stage where the provision of adequate time is more important than in design. It is cost-effective common sense.

We all know that it is far less expensive to correct things during the design stage than it is in the later stages of the project, such as during actual construction, when factors such as poor weather or site conditions, work delays, cost escalations and the need to replan activities and carry out remedial work are all likely to come into effect.

For this reason you need to apply the '*get it right from the outset*' principle to your health and safety strategy and ensure that activities such as obtaining all the required input data, timely health and safety assessments and joint health and safety reviews (for example, with the planning supervisor and principal contractor) are incorporated into your design programmes and/or networks and receive the necessary resources to implement them properly, including resources for any ensuing actions such as the distribution of information to others.

Step 4: Integrate health and safety requirements into the management system

In the previous section we discussed the need to plan and resource health and safety activities into your design programmes. You now need to go one step further and decide how you can ensure that this will happen.

The solution is to integrate those things which require health and safety activities to happen into your wider design management system.

Many successful design organizations nowadays have quality management systems which they will have established to meet the requirements of an appropriate ISO standard such as ISO 9001:2000. They will have done this in recognition of the fact that such a system properly prepared and selectively used will not only be a continuous expression of their most effective and efficient working processes, but also a vehicle through which to achieve the organization's prime business objectives – for example, profitability, return on investment or meeting fully (or exceeding) their customers'/clients' expectations. In addition, it will be a significant aid to winning business opportunities, partnering arrangements and the like.

If you want to ensure that health and safety requirements, such as risk assessments, using your database, planned/timely meetings with others and so on, become an automatic part of your design process, they need to happen by virtue of your management system's procedures. This subject is revisited in greater detail in Chapter 4.

Step 5: Communicate with others

One of the fundamentals for success in your quest for improved health and safety during construction and beyond is seen as that of communication – that is, eliminating the 'pigeonhole' mentality and, instead, working closely with others at key stages of the design process, in order to benefit from collective knowledge gained from different project backgrounds.

There are three key players who are central to this collaboration process:

- the *designer*, who attempts to identify and eliminate or reduce hazards and risks and, through the planning supervisor, makes inputs into the pre tender health and safety plan

- the *planning supervisor*, who is responsible for liaising closely with the designer, receiving site information from the client and developing a pre-tender health and safety plan to assist potential bidders for principal contractor and other roles

- the *principal contractor*, who is largely involved in developing the pre-construction health and safety plan into a full health and safety plan for construction, communicating with other contractors concerning health and safety matters and coordinating their activities with respect to health and safety.

On a sizeable project, these three parties should meet at regular intervals, during the evolution of the design, to discuss ongoing health and safety issues so that:

- the designer can brief the others on the design configurations, recognized hazards, and measures proposed or taken to eliminate or reduce the associated risks

- the principal contractor can advise or comment from a different perspective and, if necessary, make reasonable suggestions for changes to be considered

- the planning supervisor can act in a coordinator capacity and ensure that agreed actions are implemented and incorporated into the health and safety plan and health and safety file as required.

Everyone will benefit from such liaisons; it is a win–win situation.

The designer gains timely inputs from a contractor perspective and possibly learns alternative approaches for eliminating certain types of risks, some of which may be worthy of incorporating into the design health and safety database.

The principal contractor is made aware of the wider implications behind the designer's thinking and of risks that may be inherent in making changes to planned construction methods and, as a result, be in a better position to guide others and meet the obligations imposed under the various regulations, including CDM.

Finally, the planning supervisor is kept in the picture and is better able to span the designer–contractor information bridge.

An application methodology

So far, we have identified a number of potential contributory factors that could give rise to the observed failures on the part of designers to demonstrate an appropriate awareness of their obligations and responsibilities under the CDM regulations. Furthermore, we have determined that, to bring about the necessary improvements, the design organization needs to have a policy that embraces the need for health and safety in construction and beyond.

We have discussed a strategy for bringing about the realization of that policy and suggested a number of steps which, when followed, should collectively produce both a necessary health and safety culture and a vehicle for bringing about the desired goals.

The following list of sequential health and safety-related activities is not definitive, but indicative of how everything so far discussed could be brought to bear within the context of the design phase of a project.

- Receive client brief and review for accuracy and completion. Formally resolve any queries.

- Form project/design team.

- Establish project quality plan, defining scope of work, standards, responsibilities, procedures and so on.

- Advise client of their duties under the CDM regulations.

- Acquire and review all site-related information in possession of planning supervisor.

- Refer to the database checklist to see whether there are any additional factors that need to be established (or discounted) with respect to the particular site or construction.

- Discuss with the planning supervisor and arrange for the acquisition of the identified information, site visits and so forth, as indicated.

- Identify site-related hazards and potential associated risks.

- Discuss with the planning supervisor, including any proposals for elimination/reduction of hazards and risks referring to the existing health and safety database for information on known hazard types and proven options for solution.

- Prepare a conceptual/outline design in conjunction with risk assessments and provide information to the planning supervisor for inclusion in the pre-tender health and safety plan.

- Following the appointment of the principal contractor by the client, hold an initial project safety review meeting with the planning supervisor and principal contractor to discuss inherent risks and solutions intended or being considered.

- If necessary, amend design thinking to incorporate any sound proposals for eliminating or reducing risks during the construction and post-construction stages.

- Feed-back any sound ideas from the above and incorporate them into the design database for reference on future projects.

- Continue with detail design and monitor health and safety activities via periodic design review meetings. Also, hold health and safety-related meetings with the planning supervisor and principal contractor.

- Incorporate health and safety-related information into design output documents.

- Release final information.

- Plan for, and receive, ongoing feedback throughout construction. Learn lessons, improve the database and update staff training.

Health and safety, rather like quality management, is an ongoing process. You need to return to it at each stage in the process.

The above is purely a hypothetical sequence, which may be influenced significantly by information gathered or preliminary work done by an appointed designer (for example, an architect) prior to a decision to proceed and the appointment of a planning supervisor.

In the case of architectural practices there exists a standard model concerning the management of project activities – namely, the RIBA Plan of Work which assists the selection of preferred options, various 'fast-track' or 'laddering' considerations and so on.

This should obviously remain as an integral part of the architect's project managment approach, but should ideally be integrated into the wider picture. To see how this might be achieved see 'Planning, resourcing and giving transparency to health and safety requirements' in Chapter 3 and 'Managing health and safety on a project-by-project basis' in Chapter 4.

Chapter 3

Application in Design

The designer's responsibilities and limitations

In Chapter 1 we looked at a limited number (but by no means all) of the main regulations which have major significance for the construction industry and identified areas where design consider- ations could help to improve health and safety for those actually involved in site construction or post-construction activities.

In Chapter 2 we then identified five of the main considerations which come into the design equation and how they were interrelated in so far that changes to any one of them would almost certainly give rise to 'knock-on' effects for the others.

In this chapter we now attempt to bring things together in more detail, beginning with exactly who is considered to be a designer.

What is a designer?

Under CDM Regulation 2(1) a designer is described as:

... any person who carries on a trade, business or other undertaking in connection with which he:
a) prepares a design, or
b) arranges for any person under his control (including where he is an employer, any employee of his) to prepare a design relating to a structure or part of a structure.

The same regulation defines 'design' as follows:

... design in relation to any structure includes, drawings, design details, specifications, bills of quantities (including specification of articles or substances) in relation to the structure.

Using the above definitions, persons and organizations that fall into the 'designer' category would be:

- architects, design consultants or others having overall responsi- bility for design
- persons designing details of fixed plant
- contractors carrying out design duties (for example, design and build)
- persons with authority to prepare or alter design specifications for a structure
- designers of falsework or formwork
- those preparing remedial work specifications or specifying articles or substances
- interior designers.

The designer's duties and limitations

Duties

In Chapter 1 we identified the key duties which fall to the designer under CDM Regulation 13. However, with respect to the requirement that the designer avoid risks, the qualifying word 'foreseeable' is used (Regulation 13(2)(a)). This word is seen as somewhat subjective insofar that what may be foreseeable to one person may not be as readily foreseeable to another who has less, or different, experience.

Nevertheless, there is now so much readily available information on hazards and their associated risks that it may be difficult, in many cases, to argue that something was *not* foreseeable. In my opinion, this situation imposes a serious burden on designers in that they have to be almost 'all knowing'.

What, then, can you, as designer, do about this? I propose that you should take the following steps.

1 Establish a comprehensive, logically presented, readily accessible, database of hazards, risks and proven options for solution, reference to which, on a project-by-project basis, is a mandatory requirement by virtue of both the management health and safety policy and the management system procedures.

2 Discuss the foreseen hazards, risks and measures deployed to eliminate or reduce them with others, such as design colleagues, the planning supervisor and the principal contractor.

3 Give the fullest and most helpful information possible concerning any residual risks foreseeable as a consequence of the preferred design configuration.

Limitations

The various duties of the designer were discussed in Chapter 2, but it should also be recognized that there are sensible limits to those duties. For example, the designer has no obligation to:

- record the methodology whereby risks are assessed
  (It is difficult, however, to imagine that any organization which has made a formal commitment to address health and safety issues, and has in place a quality management system, not having procedures concerning both the methodology of carrying out, and the recording of the results of, such assessments.)

- provide information about hazards and risks which are unforeseeable
  (This, however, brings us back to the subjectivity of what should or could have been foreseeable and the benefits of having available comprehensive guidance information.)

- stipulate methods of construction, unless the design is based around a particular erection/construction sequence about which a contractor may need to be aware

- carry out any health and safety management role over others – for example, contractors

- review or report upon a contractor's health and safety performance
  (However, such performances would be of interest to many design organizations, such as architects, who may be asked to recommend competent organizations to their clients and who, as part of their own quality management systems, maintain a dynamic list of preferred suppliers of goods and services.)

- stifle innovation, flair and creativity

- choose the safest form of construction (for example, concrete over steel)

- possess a detailed knowledge of the construction process.

Hazards and risks in relation to design

What are hazards and risks?

A hazard can be described as *something which has the potential to cause harm or injury*. By contrast, a risk could be described as *a likelihood of harm or injury as a consequence of a hazard*.

A single hazardous situation can give rise to several risks as the following examples show:

- Working at height could result in risks of:
  - people falling during access to, whilst being at, or egressing the place of work
  - objects being dropped, causing injury to those below.

- Excavations could present risks due to:
  - collapse and entrapment of those working within the excavation
  - striking existing underground services
  - loads imposed by vehicles too near the excavation.

Common types of hazard and associated risks

In this section we list a short selection of well-known hazards, the risks that they can give rise to, plus a few pointers as food for designer thought.

In addition to the above, a number of recommended publications from which you can gain a more detailed insight into hazards, risks and measures is included in 'Relevant publications' at the end of this chapter.

Working at height

The main risk associated with working at height is that of falling, leading to injury or even death. Why, then, do people fall?

A number of reasons seem to recur. These are:

- failure of access support, such as ladders, scaffolding and the like
- ignoring the restrictions placed upon worker's movements
- workers being required to work beyond the protection provided
- lack of edge protection or inadequate edge protection
- the support structure (for example, a roof) having insufficient strength to bear a person's weight.

Always try to eliminate a risk if you can.

What can you do to help counter the above?

The first thing is to eliminate, as far as possible, the need to work at height. For example:

- Consider off-site fabrications which lend themselves to easier assembly and require less manual labour.
- Minimize the need for ladders and scaffolding by including, where possible, hard standing which could allow the use of mobile access equipment such as hydraulic platforms.
- Design stairways for use in construction.
- Use materials that don't require regular maintenance, such as repainting.
- Fit windows that can be cleaned without having to access them externally.

Where problems cannot be designed out, then 'design in' provisions to help safe access, working and egress. For example:

- Provide well-marked, safely accessible anchor points for safety nets or harness securements.
- Specify non-fragile roof sheeting capable of withstanding a person's weight.
- Include clearly marked access and movement paths for people accessing or working on roofs.
- Provide location brackets or similar to prevent ladders slipping sideways.
- Provide (or include provision for) edge protection – for example, railings.
- Specify non-slip surface materials.

See 'Relevant publications', 'Working at height' for more information.

Working in, or close to, excavations

Typical hazards associated with working in, or close, to excavations are those of:

- collapse, leading to entrapment, entombment, injury or death
- striking existing utility services
- the presence of contaminants, contacts with which would be harmful
- the build-up of dangerous gases, which could lead to explosive or poisonous atmospheres
- vehicles/equipment operating nearby causing collapse, or toppling into the excavation.

What can you do in order to help combat some of the above?

First, you must have complete and reliable site-related information. A useful, but not definitive, checklist of the typical information needed would be as follows:

- existence and location of any underground workings, water courses and services
- details of previous site usage and of any possible remaining residues
- results of geological site surveys, soil sample tests results and so on
- water table levels
- adjacent buildings and the nature and depth of their foundations.

It is important that the information required by such a checklist is established or otherwise can be confirmed as 'not relevant'. Once the above facts are established, you can then consider ways of avoiding or minimizing risks consequential to the hazard – for example, by:

Don't make assumptions about the relevance of information unless you have checked it out.

- considering the possibilities of non-excavation for service supply runs (for example, by mole-type drilling)
- situating excavations away from contaminated areas and existing underground hazards or services
- keeping excavations as shallow as is possible, by using the engineering properties of the ground itself to maximum advantage.

See 'Relevant publications', 'Working in or close to excavations', for more information.

Manual handling

Manual handling can be hazardous because of the strain which can be imposed on the human body as a result of:

- trying to handle excessive loads
- having to adopt an incorrect posture while handling a load
- supporting a load for too long
- maintaining an incorrect posture for too long
- the distance through which a load is moved
- gripping difficulties.

What can you do to counter these hazards?

- Attempt to design out the need for manual handling – for example, by providing hoists and good standing for lifting aids.
- Have early discussions with the planning supervisor and principal contractor about foreseeable construction methods and where design change may help.
- Incorporate ergonomic considerations into the design.
- Ensure that there is adequate space available for foreseeable manual handling.

See 'Relevant publications', 'Manual handling' for more information.

Asbestos

Although asbestos is a classified hazardous material, we have dealt with it separately, because it is subject to its own regulation. The dangers of asbestos are now very well documented and designers today will not (it is hoped) be specifying designs containing this material.

It is essential, however, that the principal contractor is aware of the presence of asbestos, particularly if the project in question involves the dismantling of existing structures which may contain this material.

It is also the planning supervisor's duty to ensure that any relevant information is incorporated within the pre-tender health and safety plan.

However, you, as designer, are required to make the client aware of their duties under the CDM regulations and, in particular, to provide the planning supervisor with the necessary site-related information, including any known presence of hazardous materials such as asbestos.

See 'Relevant publications', 'Asbestos', for more information.

Hazardous materials

The sources of material-related hazards on a site can be numerous. Many are often related to previous site usage which, depending on the nature of previous activities may leave residues of wastes, oil, chemicals, asbestos, and by-products from various processes, among others.

Other sources of hazard may be located within existing structures, where the possibility of releasing some hazardous substance (such as asbestos) on dismantling can be a serious risk to be considered.

Sites previously used for tipping purposes could be generating build-ups of gases or leaching chemical cocktails which may threaten existing water systems. Beyond the site itself, there may be risks resulting from converting solid materials into powder or dust form which can then become airborne and a danger to others in the proximity, or from working with materials which have low flash-points or explosive potential if close to a heat source.

You can help reduce many of the above risks in a number of ways. The following are but a few.

First, identify hazardous materials on-site as completely as possible by reference to information sources such as:

- any existing health and safety file
- site investigation reports.

Then design round the hazard if possible, using the following guidelines.

Don't forget materials can represent hazards not only in the processing, but in storage and transporting as well.

- Do not specify materials which are known to be potentially hazardous to health and safety, whether it be in use, transporting, storage or processing.
- Where possible, design for construction techniques which avoid or limit exposure to hazardous materials.
- Do not specify processes which can generate hazardous by-products (for example, cutting and chasing).
- Avoid materials which can give rise to skin irritations and diseases on contact.
- Refer to your health and safety database.
- Discuss matters with the planning supervisor and principal contractor.
- Press the client regarding their responsibilities to provide the fullest site-related information.

See 'Relevant publications', 'Harzardous materials', for more information.

Temporary works

Earlier in this chapter we identified that temporary works require equal design consideration to the main structure.

Health and safety problems associated with temporary works equipment include:

- instability and collapse
- manual handling problems, difficulties in erection
- people and objects falling from heights.

What, then, can you do to offset some of these problems?

- Design scaffolding in accordance with BS 5973.
- Avoid, wherever possible, excavations (for service runs, for example) where firm standing is required for activities such as erecting scaffolding.
- Consider how essential edge protection may be incorporated into the design, thus helping the contractor meet a lawful obligation.
- Provide tying points for scaffolding where possible.
- Ensure that falsework schemes do not demand so much space such that it impedes movement.
- Incorporate, where possible, falsework into the permanent works.
- Ensure that the permanent works are capable of carrying any loads imposed by the falsework.
- Design falsework in accordance with BS 5975.
- Advise of any forces which could cause mobile access towers to topple – for example, forces resulting from work activities, such as pulling on cables, or natural forces such as wind.

See 'Relevant publications', 'Temporary works', for more information.

Noise

Exposure to high noise levels over prolonged periods can lead to impaired hearing or even deafness.

The construction process involves many activities which can give rise to such levels – for example, shot firing, concrete drilling or cutting, pile driving, compacting and so on.

What, then, can you do to reduce noise?

The following are some areas to consider.

- Design built-in ducting for services to avoid chasing.
- Avoid drilling where possible; specify, for example, 'cast in' instead of 'drill-and-fix' type anchors.

You need to think about the work involved at each stage of the construction project to avoid exposing contractors to hazards.

- Specify that concrete blocks must be cut to size off-site under controlled conditions.

See 'Relevant publications', 'Noise', for more information.

Erecting structures

The hazards associated with structural erection are numerous, including but not limited to:

- people or objects falling from height
- temporary instability
- difficult handling of loads
- collapse of temporary works equipment.

You can play a significant role in limiting exposures to hazards during the design stage of construction. The following are some of the measures which should be considered.

- Aim to minimize erection durations by specifying suitable off-site prefabrications of handleable size and weight.

- Ensure that structural members are capable of withstanding construction forces – for example, compression forces which may be introduced during slinging.

- Provide for temporary bracing to maintain stability during construction.

- Design for accessibility and the safe use of temporary works equipment, cranes and so on if needed.

- Provide information on load-bearing capabilities of structural members in the event of circumstances which might foreseeably arise during construction – for example, how much additional load can be tolerated should temporary works equipment need to be secured to them, or their capability of withstanding forces imposed by fall from height.

- Advise on or provide within the design the securement positions of safety lines, nets and so on.

See 'Relevant publications', 'Erecting structures', for more information.

In addition to the examples given above, there are numerous other situations which present hazards and risks to health and safety and where designer input can be important. These are summarized below.

Working close to or on electrical systems

- Design provisions to isolate the system.
- Ensure that wiring diagrams and instructions carry necessary safety-related information.
- Make provision for the secondary supply of essential services.
- Plan for good lighting to aid safe site working.
- Identify any underground (or potential) supply routes, and state their depth.
- Avoid the potential for possible build-ups and discharges of static electricity.
- Include provision for lightning conduction.

Structural refurbishments (see also demolition)

Consider the possible hazards and risks associated with:

- the possibility of toxic materials
- the removal of existing materials
- new materials
- the need for temporary supports
- the containment of dust and other airborne particles
v the isolation and ventilation of work areas.

See 'Relevant publications', 'Structural refurbishments', for more information.

Confined spaces

In the case of confined spaces, such as excavations or deep basement shafts, design for:

- good access
- means of escape
- ventilation
- safe lighting
- impermeable membranes against possible leaching of gas.

Working with hand-held tools

Avoid the need for scabbling, chasing and so on.

Demolition, dismantling and decommissioning

- Consider the possibility of toxic materials.
- Specify foreseeable demolition problems such as:
  - loadings on existing structures due to demolition work (such as piles of rubble)

- safe access and egress
- maintaining a safe working environment
- planning for safe working
- needs for handling equipment
- the designation of exclusion zones – for example, 'wide' (public exposure) and 'close' (people doing the work).

See 'Relevant Publications', 'Demolition, dismantling and decommissioning', for more information.

Safe use of cranes on construction sites

See 'Relevant publications', 'Safe use of cranes on construction sites', for more information.

Working over water

- Consider prefabrication, in order to avoid fabricating over the water.
- Consider temporarily diverting the water course.
- Consider the need for a cofferdam.
- Specify appropriate personal safety equipments, such as lifebelts and lifejackets.

Figures 3.1, 3.2 and 3.3 illustrate just three examples of safe and unsafe practices in relation to typical site- and structure-related situations.

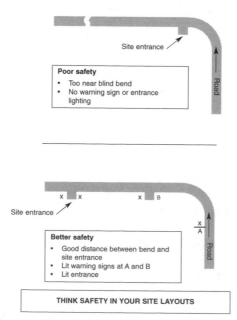

Figure 3.1 *Safe and unsafe site entrance design*

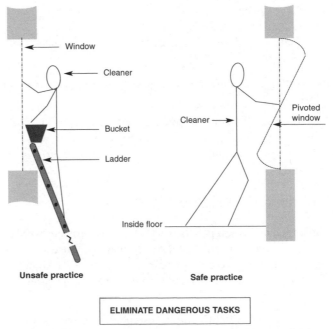

Unsafe practice

Safe practice

ELIMINATE DANGEROUS TASKS

Figure 3.2 *Eliminating the necessity for working at height*

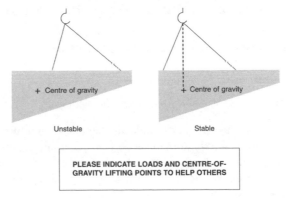

Unstable

Stable

PLEASE INDICATE LOADS AND CENTRE-OF-
GRAVITY LIFTING POINTS TO HELP OTHERS

Note: For the safe lifting of objects by slinging, the lift attachment points need to be higher than the centre of gravity of the mass being lifted and the line of lift should pass through the centre of gravity.

Figure 3.3 *Reducing the risks to others in lifting heavy loads*

Setting up a health and safety-related database

In earlier chapters we have emphasized the importance of establishing and using a comprehensive health and safety-related database.

To provide sound designs, your first need is for reliable input information. This will usually come in the forms of:

- the client's brief
- site-related information
- applicable regulations, standards and so on, which have to be complied with.

In this section we are essentially concerned with the latter two types of information.

Although the information you need will be specific to a particular project, your database and the way in which you use it, will help you to ensure that important health and safety considerations are not overlooked.

Establishing site-related information

It is important not to accept only whatever information is passed to you, if you feel more is needed. All too frequently those who are not prepared to qualify exactly what they do want, or provide all the information needed by others, will be the first to apportion blame if things go wrong.

Challenge the information you are given about a site if you feel it is complete or incorrect.

In order to help establish all the essential information it is recommended that you adopt a 'checklist' approach, making sure that this list covers every conceivable subject which could influence the design proper, including the identification of potential hazards and risks.

The following is my own conception of a rough list. It is in no way definitive and you are invited to extend it, on the basis of your and others' experiences, by using techniques such as those described in the next section.

Suggested list of useful site-related information:

- previous site usage (for example, tip, factory, agriculture)
- known residual/hazardous materials from previous usage/processes
- underground workings, watercourses and so on
- underground services – electric, gas, water, sewage, tele-communications
- streams, watercourses, ponds and other water features
- water table level
- results of soil or geological surveys
- overhead power cables and/or telephone wires
- prevailing wind direction
- limitations of site access/egress – for example, width, height, load bearing capacity, times of use
- adjacent buildings, uses, types of foundation
- demolition requirements – known hazardous substances
- preserved features and ecological constraints.
- planning and/or building constraints.

Such a list can act as an *aide mémoire* and could be laid out as shown in Figure 3.4.

Site information checklist

Project/Job Rep. No. _____

Information subject	Project applicability Yes/No	Information available Yes/No	Information source	Information obtained or still needed	Foreseeable hazard potential
1.	2.	3.	4.	5.	6.

Pre-completed list of all site-related information requirements

Completed by: _____ Position: _____ Date: _____

Figure 3.4 *Proposed layout for a site-related information checklist*

When completed and signed, the list would be then used for the next step and a copy put into the project records.

Using the checklist

The first step is to decide the relevance of the items of information listed in column 1 in relation to the project, and then write 'yes' or 'no' in column 2. For example, if there was no possibility of working over water, the answer would be 'No'; if there were overhead cables, the answer would be 'Yes'.

Next, consider whether you have the information you need. For example, if you wish to have a soil report, but do not have one, you would enter a 'Yes' in column 2 and a 'No' in column 3. In column 4 you would enter the source required to provide the report, and you would enter 'awaited' in column 5. The completion of column 6 would depend on the information produced.

Carrying out the above exercise:

- forces advance thinking about possible implications
- highlights when there is insufficient information to proceed with confidence
- flags up potential hazards
- gives visibility to health and safety considerations from the outset.

Application regulations and standards

The second type of information we identified as being needed was copies of current regulations, their associated ACOPs, national standards and so on which provide positive guidance concerning obligations to be met – for example, CDM Regulation 13 defines the duties applicable to the designer.

With regard to this type of document, we recommend using a reliable service provider who offers ongoing information update support.

Establishing information on types of hazard and risks

Another part of your database will concern known sources of hazards, their associated risks and design considerations which may help eliminate, or at least significantly reduce, the likelihoods of damage to health and safety.

Such information should be organized under hazard-type groupings. Information should be presented under key headings similar to those shown in Chapter 1 in the section on COSSH.

The actual associated risks and points to consider (insofar as may be practicable) would be identified by cross-reference to appropriate publications issued by professional bodies such as the HSE, CIRIA or the CIC from which lists and copies of their various publications can be readily obtained. Once such documents are acquired they should be identified and stored in relation to the particular hazard situation which they address.

A further dimension to the technical library/database will be any lists issued by clients, or others of appropriate standing, concerning materials not to be used under any circumstances due to their known toxic or other hazard potentials.

With regard to such materials, lists of possible safe alternatives (if known) could be helpful.

Precautions

The following should also be remembered about any database (be it computer-based or hard copy).

A health and safety database is of no value unless people are committed to its proper use.

- It is only as good as the information within it.
- It will only serve its purpose if people use it in a proper manner. (Remember what we said in the Preface: 'It is people who cause things to happen'.)
- It is dynamic and must therefore be kept up-to-date and reflective of changes in legislation, technology, new findings etc. As such it needs to be under proper controls, including the way information is accessed and used, without risks of unauthorised information copies being produced, proliferated and so on.

Additional guideline

With regard to data compilation, one approach advocated by the HSE, in its publication, *Health & Safety in Construction – Designers Can Do More?* (November 2003) is that of Red, Amber and Green lists.

- Red denotes that the item is not to be used.

- Amber denotes that the item is to be eliminated or reduced as far as possible. The inclusion of such items on the project would always necessitate the passing of information to the principal contractor.

- Green denotes that the item is positively encouraged.

The lists shown in the above-named publication should not be considered as definitive, but as merely indicative. They have not, nor should they be construed as having, been approved by the HSE.

Expanding database information using peer inputs

The collective output of a group of people is usually greater than the total of the individual outputs of those within it. This is because people react with each other, build upon the inputs of others and combine their individual knowledge and skills to reach a new joint awareness and result.

Even within a group of individuals all performing a similar task – for example, design – each player will have some knowledge or experience which is unique to them and which, if shared, would be of value to the others and indeed the organization as a whole.

Here, we discuss how we may free that individual knowledge and experience in order to expand the scope of our database for the subsequent benefit of all.

Two techniques are described below, but there are several others which could be used.

Involving people in the building of the database means that you get the widest possible source of information. It also means that they will be more likely to use it.

Extracting and developing information

I have used the following technique frequently in the past and have found it to be unfailingly successful.

- Bring together a small group of people (usually not more than six) with common experience (for example, structural designers or project managers) or common interests (for example, designer, planning supervisor, principal contractor).

- Give each a pencil and some paper and ask them to independently, and without any discussion, write down their answers to a specific question. For example:
  - 'Make a list of situations that you feel could present a hazard to those working on a construction site or to subsequent occupiers of the finished structure and, as such, merit consideration at the design stage.'

- Allow 15 minutes for this exercise and then ask each person in turn to read out their list.

Inevitably, everyone will have identified certain things in common (usually about 80 per cent consistency), and those commonly agreed items can be recorded on a master list.

There will, however, be other points that only a few or even only one person has identified, probably because that person has had an experience or has knowledge of an experience which they would always consider. These additional points can be discussed in turn and, if deemed valid, added to the master list and also to your database.

What we achieve by this simple exercise is to democratically, and within a very short timeframe, establish an information base, wider than that hitherto used by any individual which can in future be referred to by all for the betterment of health and safety in construction.

Brainstorming

This well-proven technique is a further way of developing the database.

The rules are simple and as follows:

- Bring together a number of people who are knowledgeable on the subject to be discussed.

- Have a leader who will:
  - identify in clear terms what is to be the subject of the exercise
  - explain the rules
  - ensure that the session runs smoothly
  - stop everyone talking at once
  - ensure that every point raised is recorded (a flipchart is helpful)
  - put forward ideas if others dry up
  - control events.

- Invite each person to put forward *single* points or suggestions, progressing round the group in turn and repeating the cycle. At this stage, they should not qualify or discuss these points.

- Record all points raised (however impractical they may seem).

- After everyone has exhausted their inputs, review the points listed and group similar points together.

- Discuss the points in turn, develop further if appropriate, and discard if considered trivial.

Let us assume you ask such a group to raise points that they considered to present risks to people carrying out construction work. Because their own thoughts are stimulated by the contributions of the others involved, a long list of points should quickly emerge.

An extension of this approach would be to pick the most significant risks identified from the above and taking each in turn try to identify feasible and practical ways in which they could be addressed in the design process.

Using your database to best advantage

Here we are only speaking about those parts of your database which relate to designing for the health and safety of those involved in the actual construction and post-construction, maintenance and cleaning functions. You want to be realistic and cost-effective and, in order to be so, you need to be able to apply your database information only to the degree required by the scope and nature of the project in question.

You have already addressed this need in seveeral ways, namely:

a) by use of an information checklist, as described earlier, which allows us to establish, from the outset, hazards seen as applicable on the particular project
b) by following the methodology proposed in Chapter 2 which, from the outset, allows the establishment of timely reference guidance given in the database relevant to the type of hazards identified
c) by continuing to learn and updating your database accordingly.

Performance inhibitors

The following represent what I believe are some of the main reasons that have led to the observations made by the HSE and others in relation to designers' performances to date with regard to the addressing of health and safety in construction:

Successful health and safety begins with management commitment, a clear policy and an implementation strategy.

1 lack of management commitment to really grasping the nettle and adopting a necessary health and safety policy backed up by a proper implementation strategy which is visibly management-led;

2 lack of a formal management system through which to ensure that the health and safety strategy can be made to happen and be subject to proper review;

3 as pigeonhole mentality characterized by a reluctance to meet with and learn from others who have different outlooks on construction health and safety, based on 'sharp-end' practicalities;

4 resistance to change – thinking up numerous reasons why something should not be changed when it is readily apparent that change is both logical and desirable – which, in the case of health and safety, may lead to someone sustaining a serious injury or worse;

5 reluctance to commit resources, such as people, time and money, on what some may see as a non-essential activity which can be left to others downstream;

6 people's lack of awareness concerning the significance of health and safety in the construction phase, what their personal role is and how to play it.

As can be seen, the HSE has clearly identified the shortcomings it has found in relation to the important matter of designing for safety during construction. The regulations exist and are applicable. The message is clear. *Act now!*

The benefits of teamworking and cooperation

We have already discussed a number of situations where combining the knowledge and experience of individuals within a team environment can produce beneficial results. What we haven't mentioned are the more general benefits that teamworking can produce, namely:

- a greater appreciation of critical issues
- the opportunity to learn from the knowledge and experience of others
- a greater awareness and appreciation of the roles and relative importance of others and their work
- the opportunity to contribute to the decision-making process and appreciate the benefits of teamworking
- personal satisfaction from having participated in collective achievement
- the opportunity to improve communication skills and confidence
- the opportunity to enhance one's personal image and reputation.

These, of course, are the benefits to the individual, but, as we have seen there are many benefits from this approach for the project as a whole.

Cooperation is key to achieving health and safety in the design phase. To summarize, so far we have discussed:

- techniques for developing database information, involving group participation (Chapter 2)
- liaisons between the designer, planning supervisor and the principal contractor concerning risk discussions and the 'win–win' outcome that could result (Chapter 2)
- a project design methodology in which these liaisons were included as planned activities (Chapter 2)
- various interface responsibilities between these three key players in the context of the project (Chapter 1).

Each player has a specific role to play in realizing health and safety in construction, which is a subject too important to be left to the limitations of individual experience and knowledge. The obvious way forward, then, is to have good, properly planned communication, to learn and benefit from others and together achieve a healthier and safer result.

Planning, resourcing and giving transparency to health and safety requirements

Measures to achieve the desired levels of health and safety in construction will not happen by chance or by word-of-mouth understanding. To be effective they have to take place at the right time, involve the right people, make use of the right methods and be resourced accordingly.

This means that they must be thought about in advance and visibly included in work programmes in the form of bar charts, networks and computer screen displays, and also incorporated into method statements or wider-scope planning documents such as quality plans. Failure to do this will inevitably lead to activity oversights, omissions, subsequent unplanned activities, likely project disruptions, additional costs and a greater risk to health and safety.

Health and safety information needs to be included where people will see it.

Let us return for a moment to the application methodology described in Chapter 2 and the various duties and activities we covered in our project planning.

Most design organizations (certainly those operating quality management systems) will be familiar with the concept and benefits of quality plans – both full project quality plans and detail quality plans. How might a typical *detail* quality plan (as opposed to a full project quality plan) be used to present in sequential order the activities described in Chapter 2 as well as other important information such as: activity implementing procedures, verifications, records, approvals and distributions of information to others?

The plan (illustrated in Figure 3.5) is typical and should give a little insight into the simplicity and benefits of such a document, particularly if you have in place a quality management system with the appropriate procedures. This type of plan has a number of important advantages, namely:

- It requires pre-thought about the things that need to be done.
- It acts as a live monitoring and control document, which can be reviewed during the design stage.
- It enables the right people to be involved at the right time.
- It identifies records and their timely distributions (thus avoiding a mad scramble at the end of the work programme).
- The signed-off plan, along with its other record documents, provides visible evidence of all activities being correctly completed.

Assessing risks

It is a requirement of the Management of Health and Safety at Work Regulations (see Chapter 1) that employers carry out assessments of risks to the health and safety of their employees and others arising at or from work activities.

Earlier in the chapter we identified many of the hazards than can exist on a construction site and the numerous risks that these can present. We also discussed steps that you can take foresee such

Detail Quality Plan No._____   Prepared by:_____Date:_____
Project No. _____   Position _____
(Design phase)

Completion record	Signed	Date
All drawings completed		
All verifications/approvals completed		
All distributions made		
All records compiled		

Key to referenced procedures

D1 – Contract review

D2 – Preparation of quality plans

D3 – Preparation and verification of drawings

D4 – Preparation and verification of calculations

D5 – Risk assessments

D6 – Design reviews

D7 – Conducting formal meetings

D8 – Control of technical H & S database

Key to abbreviations

C – client

PM – project manager

PS – planning supervisor

PC – principal contractor

S – self

Page 1 of 4

Figure 3.5 *Hypothetical detail quality plan, based on the sequence of health and safety-related activities listed in Chapter 2*

Number	Activity description	Control document	Action by		Verifying record	Record distribution			Comments
			Self	Client/ other		Self	Client	Other	
1	Review client brief and formally resolve any queries	Procedure D1	PM	C	Final agreed brief	S	C		Copy of brief to project file
2	Form project design team	O&R section of quality manual	PM	–	Defined team and responsibilit-ies	S			Information will be included within the project quality plan
3	Establish the project quality plan (PQP)	Procedure D2	PM	–	Completed PQP	S			Will be held in project file
4	Advise client of their duties under CDM regulations	CDM Regulation 13	PM or nominee	–	Copy of notification	S	C	PS	Copy of notification to be held in project files
5	Acquire site-related information from PS	Hazard-related checklist	PM or nominee	–	Copy of correspond-ence with PS signed-off list	S			Copy of correspondence and signed-off list to project file
6	Make initial assessment of hazards and countermeasures	Hazard checklist, database, procedure D5	PM and others	–	Risk assessment report	S		PS	Copies of report to file
7	Discuss risk and solution consideration with PS and others as appropriate	Risk assessment report from (6), procedure D7	PM	–	Minutes of meeting	S		PS	Copy of minutes to project file

Figure 3.5 *Continued*

Number	Activity description	Control document	Action by		Verifying record	Record distribution			Comments
			Self	Client/ other		Self	Client	Other	
8	Prepare design. carry out further risk assessments if necessary	Procedures D3, D4, D5, D6	Design team		Approved drawings and calculations. Design review minutes Copies of RA information to the PS	S S		As per contract PS	Copies to project file and drawings file as appropriate
9	Initiate meeting with PS and PC to discuss the design H & S arrangements and obtain input from PC	Procedure D7	PM	PS PC	Minutes of meetings	S		PS and PC	Copy of minutes to project file
10	Amend drawings if significant H & S improvements are possible	Procedure D3	Design team		Approved amended drawings	S		As per contract	Copies to project drawing files
11	Incorporate information from (9) and (10) into database as appropriate	Procedure D8	PM or Nominee		Updated database	–	–	–	–
12	Continue with detail design and periodic design reviews	Procedures D3, D6	PM and design team		Approved drawings. Minutes of design reviews	S			Copies of project file and drawings file as appropriate Page 3 of 4

Figure 3.5 *Continued*

Number	Activity description	Control document	Action by		Verifying record	Record distribution			Comments
			Self	Client/ other		Self	Client	Other	
13	Hold periodic meetings with PS and PC	Procedure D7	PM	PS PC	Minutes of meeting(s)	S		PS PC	Copies of minutes to project file
14	Check all H & S-related notations on drawings. Approve and release	Procedure D3	PM or nominee	–	Approved drawings	S		As per contr-act	Copies to project drawing files
15	Check all planned activities completed and sign off P1 of this plan	Procedure D2	PM	–	Signed off plan	S	–	–	Signed off plan to project file
									Page 4 of 4

Figure 3.5 *Concluded*

hazards and risks and to access existing information about various counteractions that may be considered.

The philosophy of carrying out a risk assessment is to consider means (insofar as they are reasonably practicable) of:

1 eliminating the risk entirely (for example, by choosing an alternative position for a structure involving excavation work, which is away from known underground workings or services)
2 reducing the risk (for example, by the provision of safety handrailing to prevent people falling from height)
3 ensuring that others who need to know (such as site contractors) are made fully aware of any residual risk situations, so that they in turn can take the necessary measures to address them.

In other words, *eliminate, reduce and advise*.

The three keywords associated with risk: eliminate, reduce and advise.

It is important to remember, at this stage, that we are not talking about risks which we all face as part of our normal living (such as slipping on an icy surface). Moreover, you are not expected to deal with the unforeseeable (although we did touch on this and how things could be made more foreseeable earlier in this chapter). Nor should you design solutions that are considered impracticable to implement (in other words, you are not expected to implement costly, resource-utilizing actions to address relatively minor problems which can almost certainly be countered by other measures).

Risk assessments can be very subjective. What one person sees as a high-risk situation may be seen by another as relatively low-risk. Questions that need to be asked are:

- What is the likely effect of the risk? Could it lead to death, severe injury or harm of a minor nature?
- What is the likelihood of the above happening? Is it proportional to numbers of people exposed to the risk, length of exposure time, working environment and so on?
- What is the likely extent of the harm? Could many people be affected or just one person?

Answers to these questions may all be somewhat subjective and, because of this, such assessments should only be carried out by individuals who are competent and knowledgeable regarding the type of activities being assessed.

This brings us to the question as to how knowledgeable the average designer is about methods of construction and the associated risk implications. (Remember it is not a requirement that they are.) The recent HSE findings give rise to concerns in this area, and it is for this reason that I would recommend that risk assessment be a team exercise involving several designers as well as the planning supervisor, if appointed. Because many of the points to be considered will be subjective, a consensus result brought about through collective inputs is likely to be a better one; at the same time, the exercise will almost certainly benefit the individual participants for the reasons we have already discussed regarding team-working.

It is important that designers do not abdicate their responsibilities and leave things to the contractor, when a little dedicated effort could make a great deal of difference.

Carrying out a risk assessment is not quite as simple as some may imagine. The following are all factors that need to be taken into account:

1 What represents the hazard?
2 What risks does it pose?
3 What is the severity of the harm that could be caused (for example, death, severe injury)?
4 What is the likelihood of its occurrence?
5 Who would be affected (for example, workers, the general public)?
6 How many people could be affected?
7 Is there likely to be any environmental impact?
8 Can we remove the hazard?
9 Can we eliminate or reduce the risk(s)?
10 Who needs to be advised concerning any residual risk(s)?
11 By whom, when and by what method is such advice to be given?

With regard to questions (1) and (2) we have already given definitions for hazards and risks in the second section of this chapter.

Questions (3) to (7) all concern somewhat subjective issues and can only be answered through the application of knowledge and experience, an open-minded approach and ideally the collective considerations and inputs of several people.

Two approaches which can prove helpful with regard to these factors are discussed in the next section.

For questions 8 and 9 you need to once again rely upon collective expertise, but in this case supported by the wealth of information in your database concerning the numerous available options.

The answer to question 10 is ultimately the principal contractor, through the initial health and safety plan prepared by the planning supervisor. (You will have already advised the planning supervisor of any residual risks so that these can be incorporated into the pre-tender health and safety plan.) With regard to the final question, it is no use developing such important information, without closing the loop. Deciding how the assessment results are to be expressed (for example, by marked-up drawings, assessment reports and/or other formal means) at what time and by whom is important. It should be made clear who has the responsibility for conveying information to the planning supervisor. The verification that it was done, by the right person, at the right time can then be confirmed during formal design reviews.

Some useful techniques for assessing risks

The numerical approach

Many organizations have adopted a numerical rating approach for carrying out risk assessments. This works in the following way.

Different hazards may put different groups at particular risk. Your risk assessment needs to reflect this.

1 An attempt is made to define who may be at risk: for example, contractor personnel, site visitors, general public, occupiers of the building, people carrying out post-completion work such as maintenance or cleaning.

    It is possible that only one group will be vulnerable in a particular circumstance, whilst several groups may be vulnerable in another circumstance, with exactly the same kind of accident. For example, in the event of a scaffolding collapsing on a secure building site, the most vulnerable would be those working on the scaffolding at the moment of the occurrence. However, an identical collapse of scaffolding that is erected on a public pavement to enable the renovation of a building could additionally lead to accidents involving pedestrians or passing traffic. It is important, therefore, that each situation be treated on merit.

2 An attempt is made to award a severity rating, giving a numerical value to different worst-case scenarios. For example:
    - Fatality: 5 points
    - Major injury: 4 points
    - Minor injury: 3 points
    - No injury: 2 points.

3 An attempt is made to put a value on the likelihood of harm being caused. For example:
    - To be expected frequently: 5 points
    - Probable occurrence: 4 points
    - Possible occurrence: 3 points
    - Not very likely: 2 points
    - Most unlikely: 1 point.

4 Consideration is also given to the environmental impact of an event (for example, pollution of a river system, airborne spread of dangerous gases and so on).

    The outcome of which will normally be categorized as:
    - high risk
    - medium risk
    - low risk.

5 The values arrived at for 'severity' and 'probably' are multiplied together and the product value used as a guide to design actions. For example, a value in excess of 10, particularly if associated with a potential environmental impact, may indicate the need to consider major redesign. A slightly lower value may indicate minor design change, if practicable, and/or advice to the planning supervisor concerning residual problems needing to be identified within the health and safety plan.

Paired comparisons

This is another group technique which is very useful for reaching a consensus decision when different group members may have individual viewpoints about a subject under discussion.

Let us imagine, as part of a risk analysis, we are considering six different risk situations and trying to decide which had potentially the most serious implications in terms of likely severity of injury and the numbers of people who could be involved.

Using the paired comparisons technique, our first step is to produce a diagram as shown in Figure 3.6 and to distribute a copy to each person.

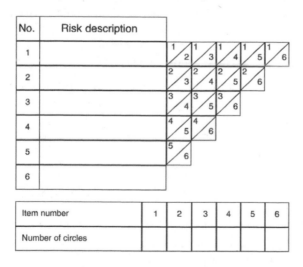

Figure 3.6 *Paired comparisons diagram*

Then list all the items under consideration (in this case, risk situations) in any order on the left. Make sure that each member of the group has the items in identical order.

Each person then independently proceeds as follows:

- Start by comparing item 1 with item 2 and circle what you feel is the most important of the two in column 1, row 1.

- Next move on horizontally and compare item 1 with item 3, again circling the one that you feel to be the most important.
- After completing row 1, continue the comparisons until every pairing has one item number circled.
- Count up the number of circles for each item and enter them in the 'totals' boxes.

Next, as a group, total up the scores for each member and you should arrive at a prioritized list. In the event of a tie, the situation can be discussed and voted upon.

The paired comparisons technique is a very good one that can obviously be used for numerous situations where some degree of prioritization is desirable.

An alternative approach

The numerical approach described earlier, although used by many organizations, is not particularly favoured by others, who consider it to be too subjective and lacking the necessary visibility as to exactly how decisions were reached.

I suggest that the following approach could counter these criticisms and simultaneously offer several significant benefits to all parties involved.

The question and answer form (Figure 3.7) leads you through a thought-provoking process, which considers risk over the full life cycle of the structure from site preparation to post-completion maintenance, cleaning and ultimate dismantling.

The questions and answers are self-explanatory. However, they are not necessarily definitive and you are invited to develop the points covered, if you feel that it would benefit your own organization to do so.

Although the methodology is described on the basis of a standard form, there is no reason why it could not be adapted for computer use. Figure 3.8 shows a specimen completed form based around a hypothetical task.

The following are seen as benefits resulting from this approach:

- It provides a pattern of thought which can be used as a basis for anyone carrying out such assessments.
- It provides visibility to the reasoning behind the decisions reached.
- It produces a visible (and auditable) record.
- It lends itself to the training of staff and the establishment of a consistent approach to risk assessments.
- Carried out as a team exercise, it enables the benefits of teamworking, as described earlier, to be realized.

1   Project/Job Ref. No.:
2   Assessment carried out by:
3   Nature of task:
4   Nature of hazard(s):

5   Foreseen risks and to whom
   a) Due to site conditions
 	
   b) During construction
 	
   c) In respect of post-completion maintenance/cleaning
 	
   d) Other (e.g. facility users/public)

6   Measures considered to eliminate/reduce identified risk situations
7   Implications of pursuing the above measure(s) in terms of practicality, resources, time and costs
8   What is/are preferred measure(s), if any?
9   What are the perceived benefits to be obtained by adopting the preferred measure(s)?
10   What, if any, residual risks will remain?
11   Suggestions for dealing with residual risks
12   How are the outcomes of this assessment to be acted upon?
13   Action implementation dates
14   Records to be retained and their location
Signed: **Date:**

Figure 3.7   *Risk assessment question and answer form*

1	**Project/Job Ref. No.:**

2 Assessment carried out by: B. Smith
  J. Jones
  B. Clever

3 Nature of task: Fixing of metal guttering and wooden fascias, soffits etc. at a height of 6 metres.

4 Nature of hazard(s): Essentially those associated with working at height.

5 Foreseen risks and to whom
  a) Due to site conditions
    None. Adequate space around the building for the erection of scaffolding or use of mobile support equipment. Good level, firm standing, no underground working or services. Clear overhead.
  b) During construction (Re erectors and others adjacent)
    • Slipping or falling whilst accessing, being at, or leaving the workplace.
    • Handling of weighty objects up to and at time of their fitment.
    • Objects falling on to others, arising from non-safe placement of tools, components and equipment.
    • Careless disposal of surplus materials.
    • Instability/collapse of working support structure.
    • Possible contact with toxic/irritant materials (e.g. paints).
  c) In respect of post-completion maintenance/cleaning
    Re maintenenace/cleaning personnel. Essentially as above.
  d) Other (e.g. facility users/public)
  e) No interface risks foreseen.

6 Measures considered to eliminate/reduce identified risk situations
  a) Use of pretreated wood and plastic coated metal sections.
  b) Use of PVC guttering, faceboards, soffits etc.
  c) Provision of gutter flushing water supply activated when required by internal means (automatic or manual).
  d) Inclusion in structural design for scaffold securements.

7 Implications of pursuing the above measure(s) in terms of practicality, resources, time and costs
  a) • would reduce rotting and corrosion
    • would reduce frequency of maintenance
    • would not reduce component weights
    • would increase costs of metallic sections.
  b) No adverse implications foreseen with respect to the above criteria (see Box 9).
  c) Slight one-off cost in respect of materials and time to install flushing pipework.
  d) Slight initial cost, but reduced erection time costs.

8 What is/are preferred measure(s), if any?
  The adoption of 6(b), (c) and (d).

9 What are the perceived benefits to be obtained by adopting the preferred measures?
  Re option 6(b):
  • Eliminates need for periodic maintenance and its associated high level work.
  • Less material/component weights, hence lower manual stress risks.
  • Lower component costs.
  • Shorter erection times and exposure to risk.
  • Less loading on main structure.
  • Less risks of impact damage if the materials/components are dropped from height.
  Re option 6(c): Eliminates the need for high-level working associated with periodic gutter cleaning.
  Re option 6(d): Safer supporting structure to work from.
  Note: The adoption of 6(b) and 6(c) eliminates post-construction, high-level maintenance and cleaning.

10 What, if any, residual risks will remain?
  Those associated with working at, accessing and egressing a high-level work place as described under 5(b).

Figure 3.8 *Specimen completed risk assessment question and answer form – continued*

<table>
<tr><td>11</td><td>**Suggestions for dealing with residual risks (see 10)**
Remind planning supervisor of the need for the contractor to provide good secure scaffolding, safety rails, toe boards and non-slip working surface, as well as the need for the safe placement of materials and tools, plus, ideally, a mechanical lifting aid and a safe means for the disposal of surplus materials from the high-level working.</td></tr>
<tr><td>12</td><td>**How are the outcomes of this assessment to be acted upon?**
By including the provisions described in 6(b), (c) and (d) in the design drawings and associated specifications.

Re 11: By meeting with planning supervisor (appropriately minuted) or by letter to the same.</td></tr>
<tr><td>13</td><td>**Action implementation dates**
Design drawings completed by: _____ (Date)

Planning supervisor advised by: _____ (Date)</td></tr>
<tr><td>14</td><td>**Records to be retained and their location**
• Copies of this assessment form and any notifications to, or minutes of meetings with the planning supervisor.
• Copies of approved drawings incorporating the features described.</td></tr>
<tr><td colspan="2">Signed: Date:</td></tr>
</table>

Figure 3.8 *Specimen completed risk assessment question and answer form – concluded*

Reviewing health and safety performance

Getting the right performance framework

How can you measure your performance? Let us begin by first looking at your 'as is' position and honestly asking yourself how well you are equipped to provide the quality of health and safety information service expected.

As your guideline use the criteria adopted by the HSE when reviewing the health and safety competence of designers and ask yourself:

- How well are your designers aware of their duties under the CDM and other regulations to eliminate or reduce, where practicable, risks to others downstream of the design function who are engaged in construction, maintenance and cleaning, and also to provide information concerning residual risks to the planning supervisor for inclusion in the health and safety plan?

- What is the quality of your risk assessment outputs? Are they factual and descriptive or merely superficial, leaving it to others to find solutions to problems?

- How many designers are aware of the numerous types of hazard that can be present during the construction phase and the measures that they can take or consider in order to reduce health and safety risks?

- Do you have a management policy backed up by procedures for health and safety in the construction phase and beyond?

- If so, is the policy and reason for it understood and are the implementing procedures adhered to?

- Do staff have a common understanding concerning the importance of health and safety and their roles in achieving it?

Be honest; don't seek to apportion blame for any shortcomings, but recognize where they exist and then determine to improve things, if necessary, by adopting an action plan similar to that described in Chapter 1.

Unless you begin by first putting your own house in order, anything subsequent is likely to be subjective to say the least.

Your performance will be a product of your degree of understanding as to what needs to be done and your commitment to doing it.

You cannot merely pay lip service to your health and safety performance.

Measuring performance

There are many criteria you can use to judge your performance with respect to health and safety in the area under discussion. The following are a few suggestions:

- Are your health and safety-related activities always visibly planned into work programmes and implemented in a timely manner?
- Are the opinions of others sought and benefited from?
- Is the health and safety information database under proper control, kept up-to-date and properly used?
- Are the outputs of risk assessments helpful and provide all the necessary information?
- Are health and safety-related features/notations being incorporated into designs – for example, provisions for safety railing, edge toe protection, lifting points, centres of gravity, safe access routes, use of off-site fabrication where beneficial and so on?
- Do design verification and review activities include cover of health and safety provisions?
- Is the inspecting authority (HSE) satisfied with your arrangements and performance?

Such questions would readily fall under the remit of an internal auditor within a design office, operating a formal quality management system. However, there is no reason within any design organization, why these questions cannot be asked within the framework of any management self-audit.

Relevant publications

Regulations

Constructing (Lifting Operations) Regulations 1961
Construction (Health, Safety and Welfare) Regulations 1996
Construction (Design and Management) Regulations 1994
Management of Health and Safety at Work Regulations 1992
Manual Handling Operators Regulations 1992
Control of Substances Hazardous to Health Regulations 1994
Any supporting ACOPs
CIRIA C604 'Work-Sector Guidance for Designers'

Use the many helpful publications to build up your health and safety database.

Further reading on hazards

Working at height

Health and Safety Executive, *Health and Safety in Construction – Can Designers Do More?*, London: HSE, December 2003.

Examples of potential hazards for designers to consider.

Construction Industry Council, Technical Guidance Notes:

TG060   *Designing to Reduce Risks Using Suspended Access Equipment*

TG070   *Designing to Reduce Risks from Working at Height*

TG071   *Designing to Reduce Risks when Working on Roofs*

T20.008   *Designing to Make Management of Risks Associated with Working at Heights Easier*

T20.009   *Designing to Make the Management of Hazards Associated with Working on Roofs Easier*

T20.014   *Designing to Make Management of Hazards Associated with Maintenance Easier: Suspended Access Equipment*

Working in, or close to, excavations

Construction Industry Council, Technical Guidance Notes:

T10.002   *Designing to Make Management of Hazards Associated with Excavations Easier*

TG040   *Designers under CDM – Designing to Reduce the Risks for Safety in Excavations*

Health and Safety Guidance series:

HSG 185   *Health and Safety in Excavations – Be Safe and Shore*

Manual handling

Construction Industry Council, Guidance for Designers, Document Nos:

H20.001   *Designing to Reduce the Potential for Musculoskeletal Injury while Constructing the Works*

HG020   *Designing to Reduce Musculoskeletal Injury*

Construction Industry Council, General Information Note, *Information on Manual Handling*.

Asbestos

Construction Industry Council, CDM Guidance for Designers Health Guidance Note, H10.002, *Designing to Make Management of Hazards Associated with Asbestos Easier*.

Hazardous materials

Construction Industry Council, Health Guidance Series:

HG010   *Designers under CDM – Designing to Reduce the Risk from Hazardous Materials*

HG10.001   *Designing to Make Management of Hazardous Materials Easier*

Other useful references listed in the above publications may also be worth following up.

Temporary works

Construction Industry Council, Technical Guidance Note: TG050, *Designing to Reduce the Risks from the Use of Temporary Works Equipment*.

Construction Industry Council, Technical Guidance Note T20.006, *Designing to Make Management of Hazards Associated with Temporary Works Equipment Easier*.

Noise

Construction Industry Council, Health Guidance Series, H20.002, *Designing to Make Management of Noise in Construction Easier*.

Erecting structures

Construction Industry Council, Technical Guidance Notes:

T20.001 *Designing to Make Management of Hazards*
and T20.002 *Associated with Erecting Steelwork Easier*

TG080 *Designing to Reduce Risks During Structural Erection*

Other documents referred within the above publications may also be worth following up.

Structural refurbishments

Construction Industry Council, Technical Guidance Document TG030.

Demolition, dismantling and decommissioning

Construction Industry Council, Technical Guidance Note TG020.

Safe use of cranes on construction sites

Construction Industry Council, Information Note 1002: *Provisions for the Safe Use of Cranes on Construction Sites.*

Chapter **4**

An Integrated Management System Approach

Incorporating health and safety requirements into the design office system

There is an increasing trend within the construction industry as a whole to establish integrated management systems embracing quality, health and safety, and environmental issues. In this case, we are assuming that your design office/architectural practice has an established management system, represented by such things as descriptions and charts of organization and responsibilities plus working procedures covering key processes/activities, backed up by standard forms and the like.

If you expect everything that we have discussed with respect to health and safety in the earlier chapters of this book to happen, you must do something to make them happen: they cannot be left to chance; they need to be part of your design system requirements.

It should be clear to everyone involved:

- what is to be done
- when
- by whom
- what evidences (that is, records) are required.

Health and safety arrangements should be an integral part of your normal management system.

Let us now reflect upon some of the main health and safety requirements that we have discussed and see how you would need to address them within any *existing* management system.

First, in terms of management responsibilities:

- You need to review your policy statement to ensure inclusions for health and safety commitment within the construction and post-construction stages.

- If necessary, you would need to establish a strategy for health and safety performance attainment.

- You need to ensure that your project O & R charts and texts make it clear who on a project design team is responsible for the carrying out of specific health and safety duties, such as:

  - reminding the client of their duties under the CDM regulations
  - liaising with the planning supervisor and others, as appropriate.

Second, you need to look at your process procedures to ensure that they adequately include for health and safety requirements. The following are seen as relevant:

- **Contract review** (that is, review prior to accepting an appointment)
  Examine the procedure to ensure that any stipulated health and

safety requirements (which may only be referred to in a very general manner, by requiring the designer to comply with CDM regulations) are fully identified, considered, understood, clarified if necessary, and can be properly resourced prior to making any commitment to proceed.

- **Project planning and resourcing**
  Ensure that any procedure addressing this subject requires that health and safety-related activities, such as risk assessments, safety reviews, meetings and communications, be included in planning documents such as networks, work programmes and quality plans.

- **Design reviews**
  Ensure that such reviews embrace health and safety expectations and verify that they have, are being or still can be met. For example, typical health and safety-related review questions may be:
  - Has timely site information been received?
  - Have necessary risk assessments been carried out?
  - Have any significant design change requirements resulted?
  - Have they been implemented?
  - What impact have they made on the programme timescale/ costs?
  - Have residual risks been formally notified to the planning supervisor?
  - Has all necessary health and safety-related information (for example, notations, handling points and so on) been included in the design?

- **Internal audits** (for those operating a QMS)
  Any procedure on this subject should ensure that health and safety measures are subject to audit.

- **Control of design base documents** (usually the technical library)
  Any procedure concerning this essential information should ensure that the health and safety-related documents mentioned earlier in this book are properly incorporated into the technical database and are subject to the appropriate controls.

- **Risk assessments**
  Make sure that risk assessments are covered by a procedure defining: objectives, methodology, implementing responsibilities, reporting findings, ensuring actions and records. The objective is to achieve a consistency of approach.

- **Preparation and use of quality plans**
  Check to ensure that any procedure on this subject requires the inclusion of health and safety-related activities – for example, liaisons/meetings, risk assessments, information distributions and so on. In the case of full project quality plans, typical inclusions would be:

- details of client, planning supervisor and other involved parties
- applicable health and safety-related design base documents such as regulations, ACOPs or standards
- liaison responsibilities on health and safety matters
- procedure references concerning health and safety activities such as the use of the health and safety database and risk assessments.

- **Training**

  Ensure that any procedure on this subject provides for staff training in health and safety matters, such as:
  - regulatory requirements
  - the importance of health and safety in construction and beyond
  - the health and safety database and its use
  - risk assessment techniques
  - designers' responsibilities, the limitations to those responsibilities, in relation to health and safety in construction.

- **Management review**

  Ensure that the continuing effectiveness of actions to meet the health and safety policy is a subject of review.

- **Quality records**

  Ensure that project records include those pertaining to health and safety activities. (Emphasize the need for this, if necessary, by explanatory staff training in any changed or new procedural requirement.)

Managing health and safety on a project-by-project basis

The key to achieving success on any project is to think ahead of the game, namely:

- What needs to be done?
- What resources (for example, skilled people, computer services, external support) will be needed to achieve the result within a given timescale and within the cost parameters?
- How will work processes/activities be sequenced, conducted and verified?
- How will your project design team be structured and who will be responsible to whom?
- Which regulations, codes and standards have to be complied with?

In other words, your first prerequisite is good planning.

Earlier in this book we have discussed many of the components which, when brought together, will help you achieve a successful result. These have included:

- training and establishing a health and safety culture
- health and safety-related regulations and guidance documents
- the establishment and use of a technical database for health and safety information
- specific key activities, such as risk assessments
- incorporating health and safety information into work programmes, networks and the like
- updating (if necessary) your QMS information – for example, O & R figures and texts and procedures – to ensure that what you want to happen with regard to health and safety is formally required to happen and can be readily reviewed, audited and monitored.

How, then, can you bring all the above into a project specific picture, which will tell you everything that is relevant concerning the particular project in question, not forgetting that your health and safety guidance and support information is already (if you have followed what has been discussed throughout the book) incorporated into your QMS?

The answer is the project quality plan (PQP) – a document with which you will probably already be familiar.

A PQP is a project-specific document (in effect a mini-management system) tailored to suit a particular project (or stage thereof) and identifying everything necessary for its management and control.

Figure 4.1 illustrates some of the key information sources which input into the PQP of a typical project, whilst Figure 4.2 illustrates, in greater detail, the fuller information that is typically included in a PQP and how it is presented.

You will notice that Figure 4.1 includes something that we have not yet discussed in this book, namely RIBA documents (such as the RIBA Job Book), which are used extensively within architectural practices. These are excellent, well-tried and tested documents containing much health and safety-related information and guidance for the profession. In themselves, however, they do not constitute a full QMS, they complement it and in those architectural practices which have established a QMS, the RIBA documents are usually integrated into it without change.

Figure 4.2 shows how everything you need to know about the project, whether for the design phase, the downstream construction phase or both, can be expressed. Within this format there is scope for using detail quality plans (DQPs) if deemed necessary. These can be used by exception if it is felt that a particular piece of work or sequence of activities (for example, a critical piece of design) needs particularly close controls. The format of a typical DQP was shown in Figure 3.5.

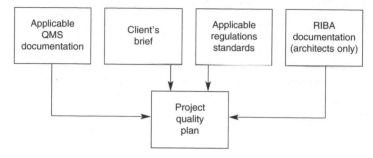

Figure 4.1 *Typical document-type inputs into a project quality plan*

Project Quality Plan No: _____			
Prepared by:	**Authorized by:**		**Date:**
Project description			
Client details			
Project team Name		Position	
External support organization/people Name	Type of support provided		
Design base data (for example, regulations, standards) Title			Issue

Page 1 of 3

Figure 4.2 *Typical project quality plan inclusions*

Project file opened?	Yes/No

Applicable management procedures
Procedure nos.

(Select as applicable from quality management system)

Other procedures to be used (for example, client-imposed)

Procedure no. and issue | Title

Applicable RIBA workstages

| A | B | C | D | E | F | G | H | J | K | L | M |

Delete stages not applicable
(Note: Concerns architectural practices only)

Authorities and utilities

Name | Contact details

Work programmes *(for example, networks, bar charts) (Use separate attachment, or if on computer give access details.)*

Page 2 of 3

Figure 4.2 *Continued*

Detailed quality plans required? **Yes/No** *(If 'yes' please describe subject(s) and scope(s).)*
Key contacts (for example, support consultants, planning supervisor, principal contractor) **Name**   **Position**   **Contact details**
Key project reviews, meetings Purpose   Timing   People involved
Inspection and test requirements (if any) Details: Page 3 of 3

Figure 4.2   *Concluded*

To summarize, the PQP identifies all the implementing information and defines the 'who', 'what', 'when' and 'how' of things, including those which concern the addressing of health and safety in construction during design and beyond.

Providing visibility of health and safety measure pre-considerations, planning, resourcing, implementation and review

The above have all been discussed to varying degrees in earlier parts of the book. Our aim in this section is to pull together some of this earlier information and see how you can demonstrate that what should have been done, has in fact been done.

One of your first duties, if the project was notifiable, was to remind the client of their duties under the CDM regulations. This is best done formally with a copy of any relevant letter being retained in the project file.

Your next health and safety consideration was to attempt to identify what hazards and associated risks were present with respect to the particular project site. You did this by using an extensive checklist of well-known hazards to see whether there was any potential hazard situation, about which you/we had no (or inadequate) information (see Figure 3.4). If you were uncertain, further information was obtained. (In carrying out this comparison you did not assume that the checklist was definitive, but rather a useful memory jogger to help avoid oversights.)

The earlier section in Chapter 3, 'Setting up a health and safety-related database', describes this early activity, the evidence (record) of which is represented by the completed and signed-off information checklist (Figure 3.4).

Having identified the potential hazards relating to the site in question, you are now in a position to undertake what would probably be the first of several risk assessments conducted at different stages of the design.

Here you use techniques such as those described in in Chapter 3 under 'Some useful techniques for assessing risks'. You would record any actions/decisions taken in order to eliminate or reduce risks (for example, amended design, revised site layout, change of materials and so on) so that the logic behind your decisions is transparent.

Your next duty is to inform the planning supervisor, providing the *fullest* information possible about the nature of any residual risks, along with as much helpful advice as you could provide for others who will be faced with the risk(s) on site or after construction (for example, during maintenance or cleaning).

This information is ideally made visible in the form of formal reports, letters and/or diagrams, copies of which are held within any health and safety section of the project file.

Finally, any joint health and safety-related meetings with others such as the planning supervisor and/or principal contractor should be formally minuted and, again, copies held in the project file.

It is important that all health and safety-related activities (which absorb people's time and cost money) are *visibly* planned into project programmes (such as networks) and are allocated adequate resources, including an allowance for possible design changes and reapprovals which may arise as a consequence of risk assessments, site feedback and so on.

Proper planning and resourcing will help your organization achieve its desired results within its time and cost projections. On the other hand, failing to adequately include those items relating to health and safety in construction and beyond will greatly increase the likelihood of programme delays and adverse cost variances during the construction phase.

As with justice, health and safety must not only be done, but also must be seen to have been done in both a proper and timely manner.

Achieving performance feedback and continuing improvement

What performance are we interested in? In the context of this book it is how well or otherwise have we, as designers, managed to overcome hazards, reduce risks or otherwise provide essential information concerning those hazards and risks to those downstream of us.

There are a number of ways of reviewing design performance.

Audits

First, you can audit your management system/processes and their application. Auditing is a requirement of ISO 9001:2000, and the technique will be very familiar to those who have a QMS in place. For those who haven't, the essence of the audit process can be described in very simple terms as follows:

- Audits are planned activities carried out by people who are trained in audit technique, familiar with the processes they are examining, but who are independent of the actual work being audited.

- Checklists are prepared. These are not definitive, but essentially *aide mémoires* to remind the auditor of the main subjects to be examined.

- The auditor, guided by the checklist, covers the range of subjects (for example, processes and procedures) to be examined, recording findings (both good and bad) concerning such matters as:
  - people's awareness of their responsibilities
  - adherence to process/procedure requirements and so on.

- At the end of the audit, the findings are reported and qualified to the management concerned, firstly orally at a closing meeting, then slightly later by means of a formal audit report.

- Corrective actions are then required to be taken within limited time periods.

- The effectiveness of the actions taken is re-examined and, hopefully, will show that the root causes of the identified problems have been remedied, recurrence has been prevented and continuing improvement has resulted.

The beginning of this chapter described how to plan out health and safety-related design activities into your management procedures and planning documents. If your audits include such documents within their scope, they will automatically check how well your health and safety requirements are being followed and actions will be taken where necessary. However, because an audit is essentially a sampling exercise and of limited scope unless it is planned to specifically target the health and safety aspects of your system, a true picture is unlikely to emerge. Something a little more focused is indicated.

Management reviews

A second technique which aims at continuing improvement is that of management reviews, although these are not frequent events and may be held only once or twice each year.

Management reviews address wider issues, such as business performance, direction, strategy, new opportunities, effectiveness of current systems and so on. They will seek to confirm that the health and safety policy is being implemented and that the corrective measures taken as a result of audits have been effective. They are not, however, the vehicle we seek, with regard to health and safety in construction.

Key performance indicators

A third and better way is to use key performance indicators (KPIs).

This approach requires defining 'measurable' criteria relating to your health and safety objectives, and then monitoring them on a planned basis, recording your findings, analysing them, identifying scopes for improvement if possible, taking the necessary measures, and then repeating the process and hopefully noting continuing improvement.

What performance indicator could you choose in respect of health and safety in the construction design phase? The following are three suggestions that relate to pre-construction activities:

1 Were site-related hazard checklists (see Figure 3.1) fully completed for each project?
2 For hazards identified was there clear evidence of design measures having been taken or considered to eliminate or reduce the risks associated with the same?
3 Were all residual hazards and/or risks fully notified to the planning supervisor for inclusion in the pre-tender health and safety plan?

Identify your own key performance criteria and use them to monitor your performance in eliminating, reducing and communicating risks.

You could collect such information after the design approvals and notifications have been made to the planning supervisor.

Similarly, information can be gathered in order to provide a picture as to how successful design health and safety considerations have proved to be during the subsequent construction stage. For example, a study over a number of projects of design change requests (which require a reason to be given for the change) will identify how many have been raised for health and safety reasons that could, or should, have been picked up during the design phase.

This is valuable feedback. The lessons learned and corrective measures taken to prevent repeat oversights – for example, by inputting information into the database or by case study awareness staff training – should be reflected in reductions of similar change requests in the future.

The KPI approach needs to be properly managed, with a named person or persons made responsible for collecting the information and reporting the findings in a properly analysed and structured form to senior management at agreed frequencies, so that the most effective measures to bring about the requirement improvement(s) can be determined.

Documentation relating to the choosing of KPIs, their monitoring, reporting, review and ensuing actions should be retained in an appropriate 'business improvements' file.

Finally, the effectiveness of the KPI approach can be monitored by your management reviews.

Chapter 5

Conclusion

In this book we have travelled a considerable distance along the road towards designing for site safety and beyond. In doing so, we have touched on many issues, not least the problems that have been identified by the HSE and others.

- We have explored the contributing reasons behind such problems and how we can address their root causes by having a sound health and safety culture and implementation strategy within our design organization.

- We have looked in broad terms at the design function and the perceived need to widen its health and safety focus beyond functional and structural safety to also more fully embrace the actual construction phase and beyond.

- We have learned a little about a selection of the numerous regulations which demand attention to these particular areas of health and safety and how the designer can, and should, play a significant role in improving health and safety for those involved in downstream activities.

- We have discussed a number of the more familiar types of hazard that can affect construction activities, and have also identified information which can not only provide helpful guidance concerning the addressing of these hazards and their associated risks, but could also form part of an integrated health and safety database.

- We have discussed the importance of collaboration with others and the potential health and safety benefits that may be derived through joint discussions with others who may see common health and safety issues from a slightly different perspective. We have also looked at the wider benefits to be gained from teamworking.

- We have established how provisions for health and safety should be visibly incorporated into our project-related planning, resourcing, programming, implementation and review activities and indeed into our design management system proper.

- We have gone beyond the current picture and considered measures that will help us ensure that our health and safety performances are not only fulfilling our current expectations, but will also continue to do so through ongoing improvement.

In the months and years ahead some of the health and safety regulations we have referred to are likely to be amalgamated or replaced. This is the nature of learning from experience and progress. What we use as our benchmarks today are merely snapshots at a moment in time.

Regardless, however, of what changes may occur, certain fundamentals will remain – namely the needs for:

- people awareness and commitment concerning health and safety
- proper management-led health and safety policies supported by clear attainment strategies implemented through management system requirements
- good cooperation, communication, knowledge-sharing and the appreciation of others and their work
- good reference data
- continuing performance monitoring and improvement.

These things will not change, and they fundamentally represent what this book has essentially been about.

If only one fatal accident or serious injury is prevented on a construction site, because somewhere up the line, a designer did a little bit extra because of what he or she has read in this book, then writing it will have been worthwhile.

Index

ACOPs (Approved Codes of Practice) 54
application methodology 36–7
asbestos 45
    publications 75
audits 29, 79, 86–7

brainstorming 57

checklists 52–4
clients, duties 17–20
communication 8, 35–6
confined spaces 49
Construction (Design and Management)
      Regulations 1994 (CDM) 15–24
    key players 16–24
    scope 15–16
Construction (Health, Safety and Welfare)
      Regulations 1996 (CHSWR) 10–15
construction work, definition of 10–11
continuing improvement 86–8
contract review 78–9
contractors, duties 22
Control of Substances Hazardous to Health
      (COSHH) 8–10
cranes, safe use of, publications 76
culture 29–30

databases 55
    expanding information 56–7
    hazards 9–10, 54–5
    risk assessment 5–6
    risks 54–5
    setting up 31–3, 51–5
    use of 57–8
demolition 49–50
    publications 76
design
    application methodology 36–7
    definition of 40
    documents 79
    hazards 42–50
    risks 42–50
    substandard 26–8
design office 78–80
design programmes 33–4
design reviews 79
designers
    definition of 40
    duties 22–3, 41
    limitations 41–2

responsibilities 26
detail quality plans (DQPs) 84–5

electrical systems 49
emergency facilities 7–8
employee information 8
excavations 44
    publications 74

hand-held tools 49
hazardous materials 8–10, 46, 55
    publications 75
hazards 54–5
    definition of 42
    publications 74–6
health and safety
    assistance 7
    culture 29–30
    file 19–20
    liaison 6
    management of 5–8
    procedures 6
    systems 6
    visibility of compliance with obligations 85–6
Health and Safety at Work, etc Act 1974
      (HASAWA) 3–5
    codes of practice 4
    guidance notes 5
    regulations 4
Health and Safety Executive (HSE), audits 29
health and safety inspector, role and responsibilities
      23–4
health surveillance 6–7
height, working at 43–4, 51
    publications 74

internal audits 79
ISO standards, 9001:2000 34, 86

key performance indicators (KPIs) 87–8

liaison meetings 35–6
lifting heavy loads 51
Management of Health and Safety at Work
      Regulations 1999 (MHSWR) 5–8
management review 80, 87
management systems 34, 78–80
manual handling 10, 45
    publications 75
Manual Handling Operations 1992 10

method statements 60

noise 47–8
    publications 75

peer inputs 56–7
performance
    feedback 86–8
    frameworks 72–3
    inhibitors 58
    measurement 73
planning 80–81
planning documents 60
planning supervisor
    appointment 17
    duties 20
    provision of site information 18
policy statements 30
post-construction problems 32–3
principal contractor
    appointment 17–18
    duties 20–21
    health and safety plan 18–19
project planning 79
project quality plans (PQPs) 81–5
project resourcing 79

quality plans 60, 61–4, 79–80, 81–5
quality records 80

red, amber and green lists 55
regulations 54, 74
RIBA documents 81, 85
risk assessment 5–6, 60, 65–7, 79
    numerical approach 67–8
    paired comparisons 68–9
    question and answer form 69–72
    records 6
risks 54–5
    definition of 42

site entrances 50
site information
    establishing 52–4
    provision to planning supervisor 18
standards 54
structural erection 48
    publications 75–6
structural refurbishments 49
    publications 76
structure, definition of 11, 16

teamworking 59
technical library 79
temporary works 47
    publications 75
training 31, 80

water, working over 50
work programmes 60